Sharp Series

Early Math Concepts Volume 1

Enrichment
Conceptual Learning

Lex Sharp

Fields of Code Inc.
Calgary, Alberta
www.fieldsofcode.com

Copyright © 2017 Fields of Code Inc. All Rights Reserved.

Published by Lex Sharp
Fields of Code Inc.
Calgary, Alberta
Canada
www.fieldsofcode.com

No part of this publication may be reproduced in any form or by any means, including scanning, photocopying, or otherwise without prior permission of the copyright holder.

ISBN: 978-1-520-581200

FIRST EDITION

Table of Contents

Prologue 1
- Conceptual Learning 1
- Interactive Content 1
- Vocabulary 1
- Third-Party References 2

Chapter 1: Set Theory 3
- Why Venn Diagrams? 3
  - Conclusion 11
  - Examples 11
- Operations on Sets 13
  - The XOR operator 24
  - XOR-ing Three Sets May not be What You Expect ... 25
- Uses of Set Theory 29
- Take a Quiz on the Set Theory Chapter 30

Chapter 2: Fractals 32
- Types of Fractals 33
- Self-Similar Fractals 34
- Non-Self Similar Fractals 38
- Take a Quiz on the Fractals Section 40
- The Menger Sponge 41
  - Conclusions 45
  - Volume and Surface Area 49
  - Surface Area of the Cube 51
  - Increasing the Surface Area of the Cube 52

    Usefulness of the Menger Sponge Model 56

    Take a Quiz on the Menger Sponge Section 56

  The Sierpinski Triangle 58

    Area and Perimeter of the Sierpinski Triangle 62

  The Sierpinski Tetrahedron 65

    Surface Area and Volume of the Sierpinski Tetrahedron 68

  Take a Quiz on the Two Sierpinski Algorithms 72

  Make Your Own Self-Similar Fractal, Create an Algorithm 74

  Examining Algorithms Further… 75

Chapter 3: The Fibonacci Sequence and the Golden Ratio 79

  Building the Equivalent Numerical Sequence... 86

  Building the Fibonacci Spiral 89

    Test your understanding… 90

  Application of Fibonacci in Sciences 91

  Take a Quiz on the Fibonacci Topics.. 91

Chapter 4: Knot Theory 93

  Trefoil Knots 93

  Borromean Rings 93

  Applications of Knot Theory 98

  Topology 98

  Take a Quiz on the Knot Theory 99

Chapter 5: The Mobius Strip 101

  The Mobius Band in the Real World 108

  Take a Quiz on the Mobius Strip 109

Vocabulary 110

- 3D and 2D 110
- Adjacent 111
- Algorithm 111
- Circumference 111
- Clockwise and Counterclockwise 112
- Consecutive 113
- Cube 113
- Equilateral 114
- Finite 114
- Infinite 114
- Intersection 115
- Microscopic 118
- Paradoxical 118
- Perimeter 119
- Perpendicular 119
- Pyramid 120
- Ratio 121
- Spiral 123
- Tetrahedron 124

Errata and Feedback 126

Prologue

This book is designed to be a Mathematical enrichment for ages 8 to 18. There are no prerequisites. The book was compiled to foster the love of Mathematics, and introduce terms not encountered routinely.

An effort was made to translate several concepts into their graphical visual counterparts, like The Fibonacci Sequence and the Golden Ratio. The goal is to lay the foundation for abstract thinking and evolve an ability to interpret visual elements and design diagrams.

Conceptual Learning

Unlike traditional learning, conceptual learning focuses on a deeper understanding of concepts and binding them into a connective mental map that generate more associations later. The goal is then for the students to realize broader concepts on their own.

The material in this book encourages an autodidactic experience, logical thinking, and conceptual learning.

Interactive Content

This volume's webpage can be found at:
http://www.sharpseries.ca/em/r1.html.

Each chapter ends with an online self-grading quiz that tests the student's understanding of the subject. Quizzes are anonymous and there is no limit to the number of times responses can be submitted.

Vocabulary

Vocabulary is the first stepping stone in academic success. Students who have a vast vocabulary ahead of time, are sure to enjoy reading earlier because the stories and articles they are reading are making sense.

To fully enjoy Math, one needs familiarity with a multitude of terms. Starting as early as kindergarten is possible, but don't be reluctant to catch up at any age.

This book has a vocabulary section at the end. Terms styled as hyperlinks have entries in the vocabulary section so students can review the meaning.

Third-Party References

This volume references unaffiliated third-party materials such as YouTube videos, and other websites. These were added to show university, research, or children's projects that were interesting and relevant to the topic. One such that should not be missed is a group of [MIT students building of a Mega Menger Sponge](). This is a project a Math club may really enjoy doing with a large group of children.

At the time this book was written, all links were checked for age appropriate content.

It is impossible to predict if these links will remain active over time. However, since they are only part of an extended framework and bonus material for further discovery, these will not affect the core content of the book even if removed.

This writer had no part whatsoever in authoring the third-party references.

Chapter 1: Set Theory

Set theory is about finding commonalities and differences between groups of things. It is part of a branch of Mathematics called **Logic**. Groups are typically called sets, and the objects that belong to the groups are called the elements or members of the set.

Set Theory often uses **Venn Diagrams** which are visual ways to describe the relationships between elements of a grouping strategy. A **grouping strategy** is the idea used to sort selected items.

Here is an example of a grouping strategy: sort all buttons in a way that distinguishes those with 4 holes from the ones that are brown, and separate them from all other buttons that are neither.

Venn diagrams are made of:

- items to organize,
- a collection of circles to organize them in,
- the constraints used to arrange the items into the circles of the diagram.

Venn Diagram circles are always shown to intersect. The intersection may or may not have elements in it.

Why Venn Diagrams?

You are probably wondering why use circles to sort things. Is this the best way to examine relationships between categorized and sorted objects?

A few attempts at categorizing elements will clarify.

The following is an example of buttons to organize based on the grouping strategy: *sort all buttons in a way that distinguishes those with 4 holes from the ones that are brown, and separate them from all other buttons that are neither.*

Examining the given set, it appears that some buttons have 4 holes and some have 2. There are several colors but the focus is on the brown buttons.

Sorting can be based on these steps:

1. place buttons that have 4 holes in one group,
2. place brown buttons in another group,
3. everything else will be separated from the rest in a third group.

The steps used must adopt a way that makes the sorting visible. Initially, we'll use arrows to point to elements and change the color of the arrows when switching to a different category.

The 4-hole buttons are identified with red arrows. The image is already quite busy and the sorting has barely started.

Additional arrows must be placed. We need a new color: blue arrows will point to all the brown buttons.

This method is overwhelming and not visually appealing. There are too many arrows. When sorting a larger number of buttons, more arrows are needed and they can get more difficult to place, especially in the middle of the image.

Let's try another method. Instead of arrows, we'll surround the group of buttons having 4 holes with an outline (left).

This is not visually appealing either. Shading (right) could be used, but makes no difference, the outline is still messy. It's going to be even messier when adding a second overlapping contour for the brown buttons.

Instead, a simpler way is to separate items in distinct piles. Let's call them A, B, and C. First, all 4-holes buttons are placed in pile B. Then, all remaining brown buttons are placed in pile A, the rest should then be neither brown nor have 4 holes. We'll call this pile C.

This looks tidy, but a problem occurred. The group of browns (set A) is missing a few items because pile B was built first. The only care for pile B was that buttons should have 4 holes. All buttons were acceptable the color was no concern so brown buttons were included. After fixing the issue by moving all brown buttons to pile A (as shown), a new problem emerged.

Early Math Concepts, Volume 1

Now it is pile B that is missing several buttons with 4 holes. The missing buttons are the same ones that were just moved to pile A.

It seems whenever pile A is fixed, pile B is ruined, and the other way around. Sharing between piles A and B would be handy.

A possible solution is using Venn Diagrams because sharing is allowed.

Let's focus only on piles A and B for now. These will be represented by two circles: red for pile A and blue for pile B, overlapping in the middle, in yellow, provides a place for sharing. The overlap is called an intersection.

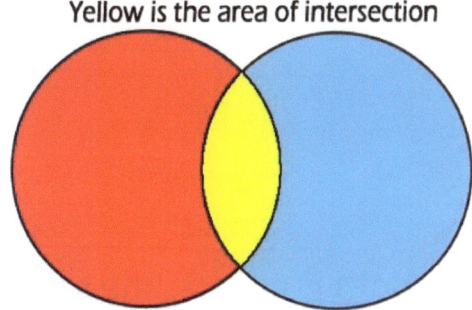

After setting up the intersecting circles of the Venn Diagrams, buttons are assigned to each area as follows:

1. The buttons that caused the problem are placed in the intersection of the two circles.

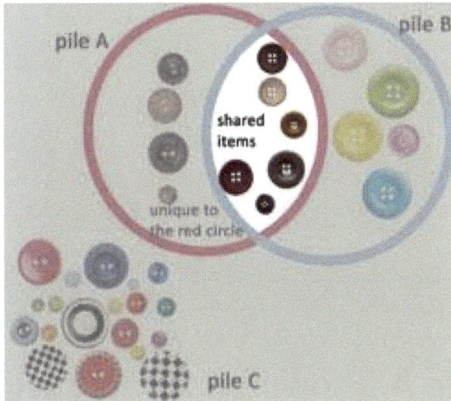

2. Buttons that are brown but do not have 4 holes are placed in pile/circle A on the left, avoiding the intersection zone.

7

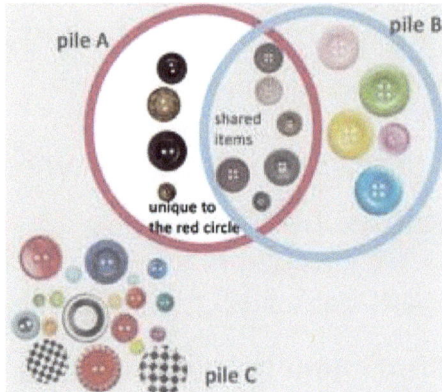

3. Buttons that have 4 holes but are not brown are placed in pile/circle B on the right, avoiding the intersection zone again.

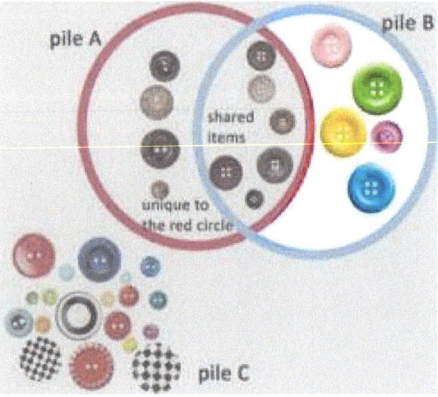

In this scheme, the problem buttons belong to both groups at the same time and are no longer an issue. The intersection is part of both groups A and B, none of these misses any of the buttons they must contain.

The final arrangement is shown below.

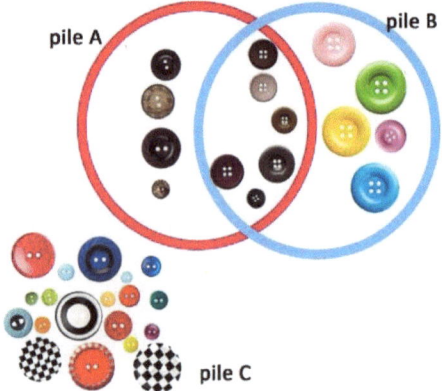

Let's review the diagrams starting with **pile A**, shown on the left below. The red circle has more elements than pile A had in the earlier methods: its unique elements plus the ones shared in the middle. Buttons appear either on the left side of the circle, as unshared unique elements in the mid, shared area.

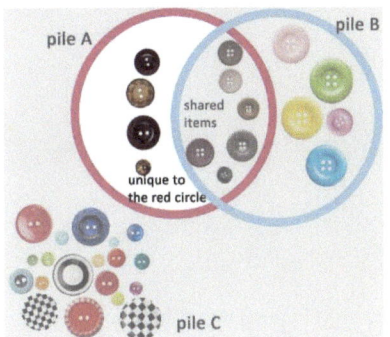 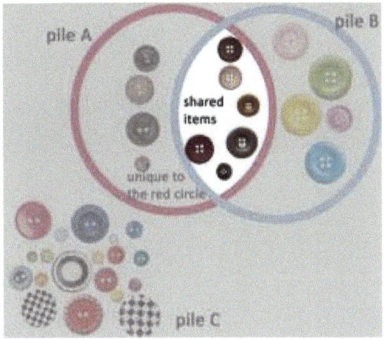

As for **pile B**, shown on the right below, the blue circle also has more elements now than it had in the previous methods. Unique elements are on the right, and buttons that are both brown and have 4 holes are shared in the middle section.

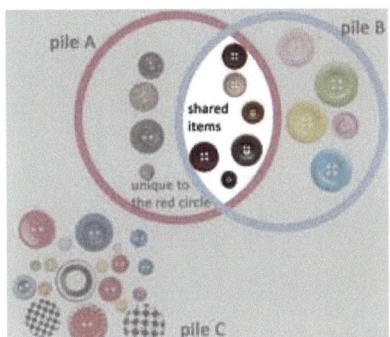 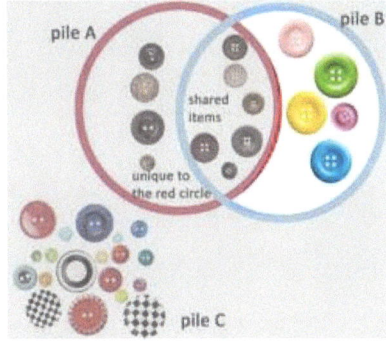

All other remaining buttons are moved into a separate pile that does not participate: **pile C**.

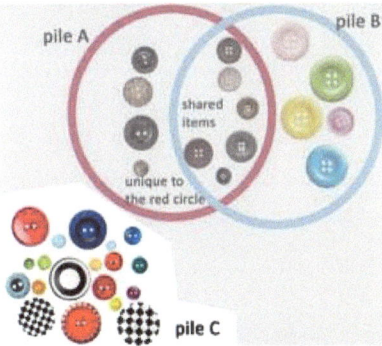

The most important part for the Venn Diagram to work, is to have the two circles overlap on the challenging buttons (see below). This is the [intersection](), where items are shared equally.

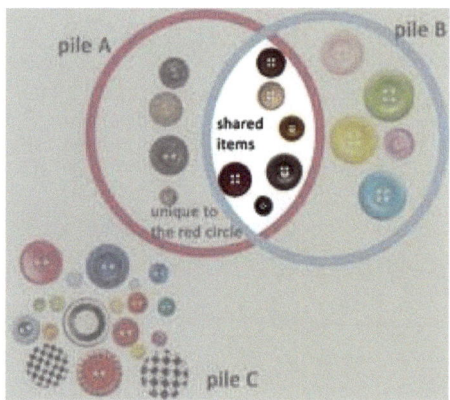

To be formal about the Venn Diagram, a third circle containing the items from pile C must be added. These are the items that were neither brown nor had 4 holes, and thus were excluded from the sort. This group has no common traits with the other two circles, their intersection is empty.

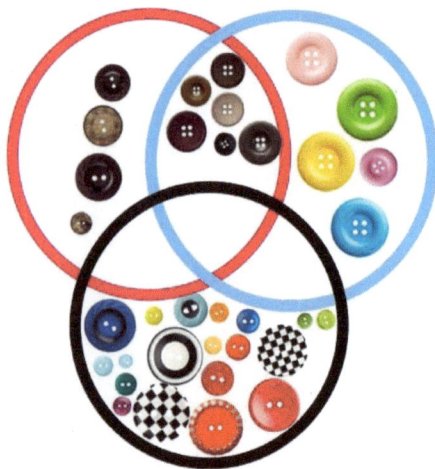

Venn Diagram are visual with easy to spot shared items, and empty space for no commonalities, like the ones pointed by the arrows below.

Visual representations are efficient. Compare them to an example of a wordy interpretation of the same content:

- The red arrow shows there is nothing in common between
    - the black circle, and
    - the group of buttons that is brown, inside the red circle.

  Therefore, the overlap of these two intersecting areas is empty.

- The blue arrow shows there is nothing in common between
    - the black circle, and
    - the group of buttons that have 4 holes.

  Therefore, the overlap of these two intersecting areas is empty.

- The black arrow shows there is nothing in common between
    - the black circle, and
    - the group of buttons that are common to both brown and 4-holes buttons.

  Therefore, the overlap of these two intersecting areas is empty.

Conclusion

Venn Diagrams provide an easy solution to show ownership over elements that are shared by several groups. They are considered powerful because they visualize otherwise verbose descriptions in a compact structure.

Examples

In the previous instance, elements had attributes that had something in common. Several buttons were both brown and had 4 holes. Not all organizing strategies are such. Sometimes sets have nothing in common at all.

Let's say a collection of toy robots must be sorted on paper circles based on these rules:

1. the red circle must contain only happy robots, and
2. the blue circle must contain only sad robots.

The arrangement is shown below.

In this example, there are no commonalities between the two circles. This is because the arrangement is based on a mood which cannot be both happy and sad at the same time. Such a state is called "**mutually exclusive**", since one choice excludes the possibility of the other. Sharing items across sets based on that idea was possible for buttons, but not so for the robots' moods.

The robots in this figure were placed in circles as required for Venn Diagrams, but the circles were disconnected and placed side by side without "interacting". Remember this model does not form a proper Venn Diagram.

To fix this, the circles must overlap in an intersection, whether common items are present or not. The result is shown below.

"**The two sets not** intersecting" fact is thus made visually obvious, and the proper structure of Venn Diagram is complete.

These quick visual cues are the hallmarks of why Venn Diagrams are so useful.

Every diagram is an arrangement based on a specific idea to sort by, like the robots' moods.

If you change the idea behind the sort strategy, you must change the resulting sets as well. The new sort will have its own combination of items for each circle. If the new sort results have common items, they will overlap in the middle and will no longer be empty.

Operations on Sets

Mathematicians like to use names for Venn circles that are simply letters. This helps them write mathematical sentences that express what is happening in a precise way. Here are a few such sentences and their translations.

Expression	Meaning
$A \cap B = ...$	The **intersection** of sets A and B equals ...
$A \cup B = ...$	The **union** of sets A and B equals ...

A ⊂ B	Set A can be found inside set B. Otherwise known as a subset. This represents an **inclusion** relationship between the sets.
A ⊆ B	Not only that set A can be found or included inside set B, but it's possible the two sets are equal. The line at the bottom is the bottom half of an equal sign. The wording for this symbol is "**included or equal**".
A − B	The subtraction of sets shown removes all elements from A that also belong to B. In other words, it deletes the intersection **A ∩ B** from A.
{red, green, blue}	When elements of a set are named one by one, it is done inside curly brackets. A list in curly brackets is a way to say the list is treated as **a set** of the "Set Theory" kind.
Sad robot ∈ A	Sad robot **belongs to** set A.
Sad robot ∉ A	Sad robot **does not belong** to set A. The symbol is the same as above but crossed out indicating the belonging does not occur.
Empty	**Empty** is noted as the result of some expressions. For example, the intersection of sets of robots that are both sad and happy is reported as **Empty**. **Empty** is sometimes referred to as **NULL** instead. In other cases, the word **Nothing** is used. Zero is almost never used to show the **Nothing** of Set Theory, because it is easily confused with the actual number zero, which can be a regular item in a set of numbers.

Some of the symbols listed above are operators similar to the Mathematical symbols **+**, **-**, **X** and **÷**. They often behave the same way, though not all of them. While intersections have no relation to one familiar arithmetic operator, unions are just like additions, and using minus for removing sets is somewhat similar to subtraction.

Let's look at how a few of these set operators work.

Names, i.e. letters are needed for each circle. The blue circle will be A, the red one B.

Then for these two sets it can be said that:

A ∩ B = Nothing

Or

A ∩ B = Null

Let's try a different sort strategy from here to practice intersections that are not empty, use these rules:

1. robots that have blue parts go in circle A (blue),
2. robots that have red parts go in circle B (red),
3. all others that have neither blue nor red, are placed in circle C.

The interesting part is to pay attention to what happens to the toys that have both blue and red on. The intersection of the circles is expected to have items with both colors.

The following is a mathematical expression of the intersection.

A ∩ B =

Construct the Venn Diagram according to the three given rules above.

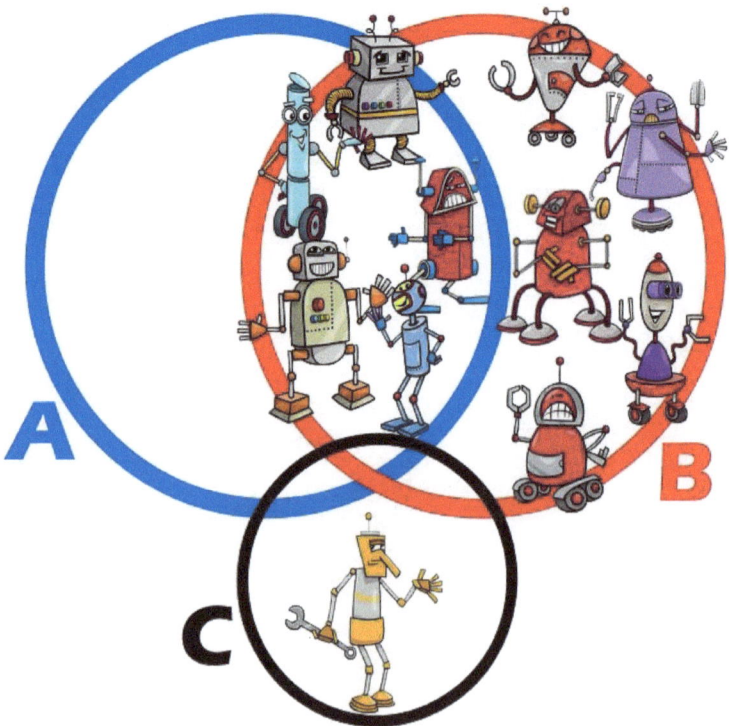

If we were to color code the arrangements we would get the following zones:

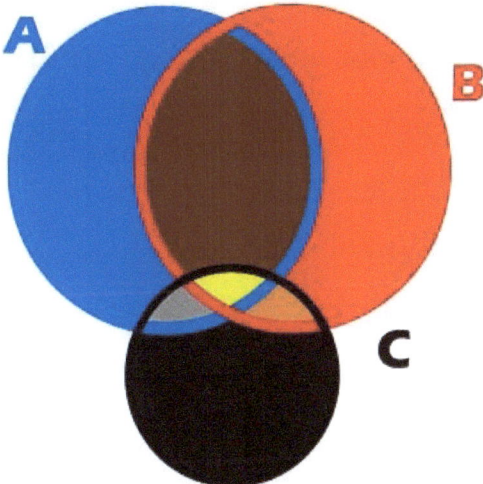

Red – contains robots that have only red and no blue on.

Blue - robots that have only blue and no red on. In this example there are none, as all robots with blue also have red on them.

Brown – the intersection between blue and red, only robots that have both blue and red on them should be placed there.

Black – robots that have no red or blue on them at all.

Grey – it's an area that does not make sense for this specific sorting order. For this section to have any robots in it, they would have to have blue on, so they can belong to the blue circle, but at the same time they cannot contain any blue at all, so they can belong to the black circle. This is a paradoxical and nonsensical requirement, and thus the area should be left empty.

Orange – similar to the grey area. It's an intersection that does not make sense in the current sorting order. For it to have robots in it, they would have to have red on, so they can belong to the red circle, but at the same time they cannot contain any red at all, so they can belong to the black circle. This is a paradoxical and nonsensical requirement as well, and thus this area should be left empty too.

Yellow – is an area identical to the grey and orange, that would expect robot to have both blue and red, but no blue and no red at all paradoxically. This is a requirement that is impossible to fulfill. Nothing can go in this zone, and nothing can represent this idea other than empty space.

Examining the diagrams below shows that the entire set of A is included in the B set. A has no items outside its intersection with B. The mathematical expression that conveys this is written as: **A ⊂ B**. We didn't use **A ⊆ B** because A is not equal to B. B has its own additional items and is thus different.

Other operations on these sets are possible, such as unions. The symbol for union is **∪**, a typical expression looks like this:

A ∪ B = {some set elements separated by commas}.

Union behaves just like the Arithmetic operator plus (**+**). A union happens when all the elements of two or more sets are bundled, or added together into one larger set. For example, the union of **A ∪ B**:

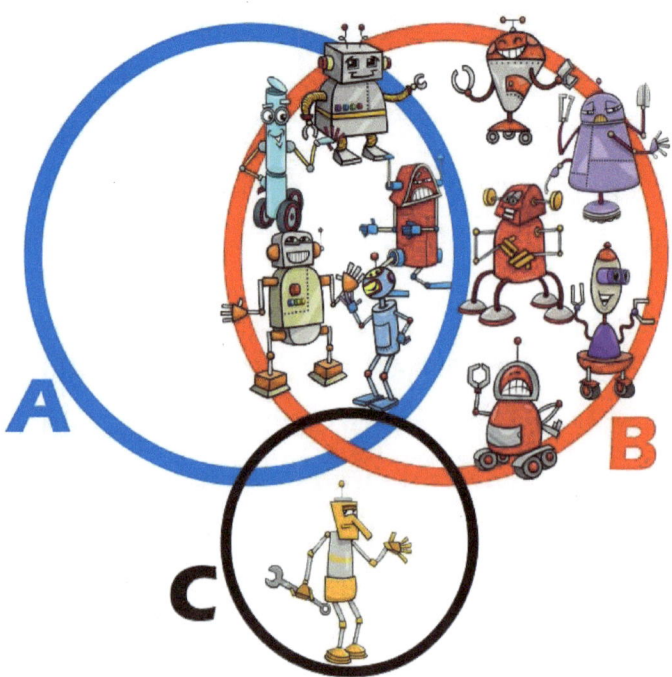

will place all items of A and B in one larger unifying diagram, shown in purple below.

Notice items of the group C are still apart in their own bubble because the operation was not **A ∪ B ∪ C**, but rather only **A ∪ B**.

Subtracting works similarly to how it does on numbers in Arithmetic. The symbol for subtracting sets is minus, for example: **A - B**.

There are a few differences when comparing subtraction of sets to the subtraction of numbers.

To clarify, we'll examine **A - B** versus **B - A**. The focus for sets is to remove items one by one until the entire subtracted group is gone.

Let's assume our starting sets before applying the minus, are the same as before.

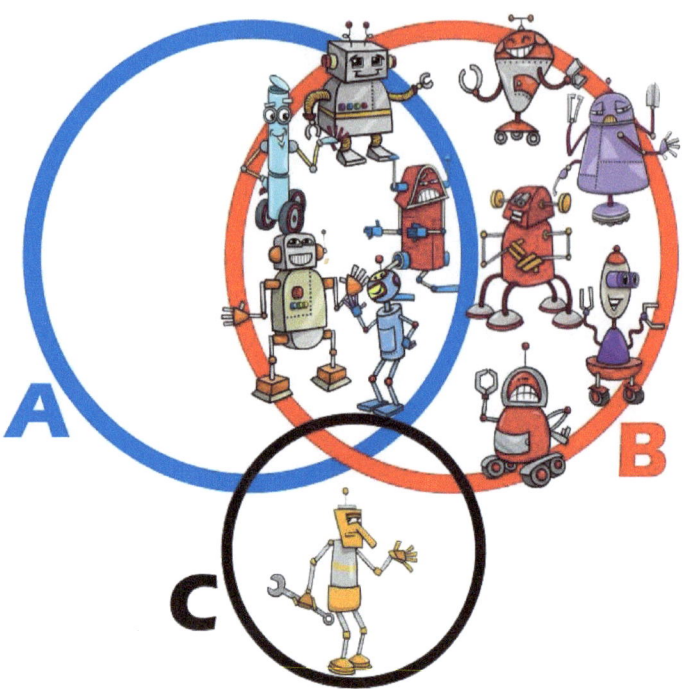

In this example **A - B** means we are removing all elements of set B from set A. However, only elements of B that are actually present in A can be removed, that is you can never get a negative set. If you were to cross out all matching elements of B from A you would get this arrangement:

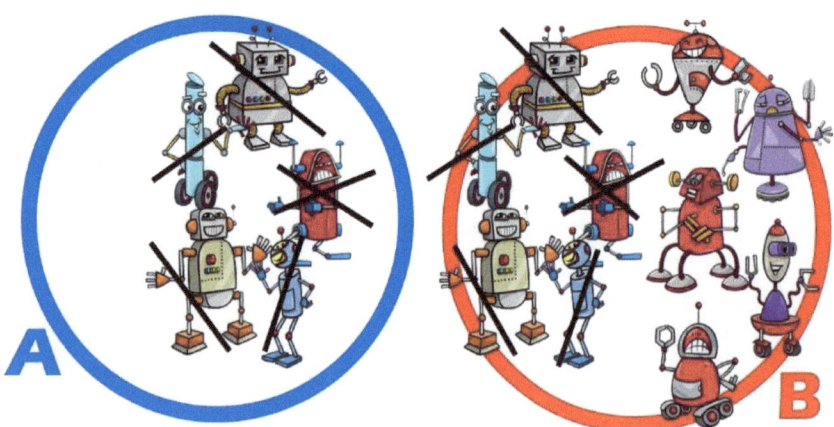

This is equivalent to removing the parts of the intersection. Practice crossing out your own sets using pen and paper to experience how this works.

After removing the intersection items of B, no elements are left in A.

Therefore, **A - B** = Nothing.

Let's also test the opposite direction, **B - A** which translates into: remove all elements of A from set B. This means removing only the elements of the intersection, and since those are the only parts of A that are present in B, the result contains pieces of B only, and none of A, as shown below. Thus, removing elements of A will not create a negative set. Examining the visual version is a lot easier than tracking a verbose description.

It is important to notice that subtracting sets differs from the regular subtraction of numbers in Arithmetic. The relationship between (A – B) and (B - A) is different than it is for numbers. It doesn't result in identical numbers with an opposite sign.

To clarify the difference, in Arithmetic:

4 – 3 = 1, and

3 – 4 = -1

For numbers, the result of swapping the positions of A and B around the minus sign (A – B) and (B - A) results in the identical numbers with opposite signs, that is 1 and -1 in the example above.

Subtracted sets do not behave this way.

Compare set results side by side. We start with:

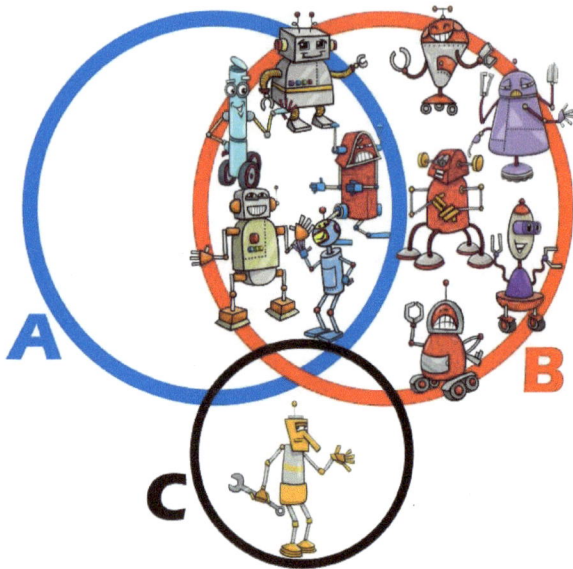

Then apply the subtractions in both directions: from B on the left, and from A on the right.

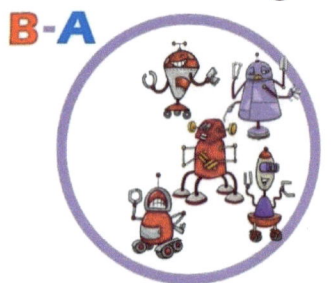
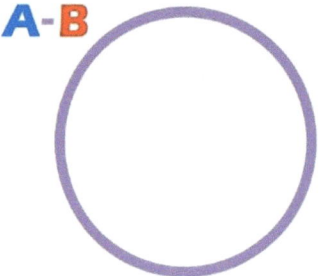

In other words:

B − A = { 🤖, 🤖, 🤖, 🤖, 🤖 }

A − B = Nothing

Set subtractions depend on the number of unique elements in each set. Subtraction of sets are often referred to as set **complements**. The wording is used as follows:

B − A is the **complement of** A in the union of B and A (i.e. **B ∪ A**). An additional notation is available for this: **B \ A** which denotes the complement. It is notable that the complement of a set can exist only in relation to a second set, never by itself, and is worded: **B \ A** meaning "the relative complement of B with respect to a set A".

In other words,

B \ A =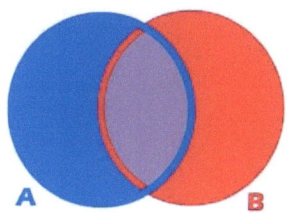

A \ B =   **Null**

We have seen what these complements look like when listing elements one by one. Let's see what they look like as geometrical shapes.

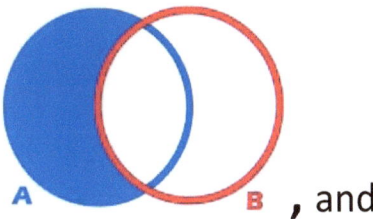

After removing the necessary areas to construct the complement (a.k.a. the subtraction) we are left with:

A − B = A \ B, which is graphically equivalent to

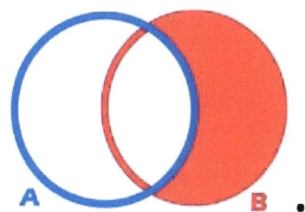 , and

B − A = B \ A, which is graphically equivalent to

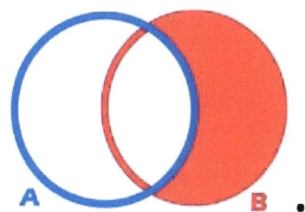 .

Notice in both cases A\B and B\A, the A ∩ B intersection (purple) is always missing from the complements.

The XOR operator

The Set Theory operators listed so far were basic. The operator **XOR** (pronounced k'sor) is an example that is more complex. It is compound, built on simple operators used in a particular order.

XOR's symbol is ⊕. The XOR of two sets is equal to whatever is left after creating a union of the two sets and then removing their intersection. Let's explore this idea for two sets A and B.

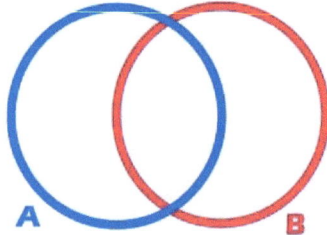

Rather than working with two empty circles, we'll emphasize the major zones with color. Three colors were used:

- blue to show all the unique parts of A,
- red for all the unique parts of B, and
- purple for the blended area where blue and red mix, a.k.a. the intersection.

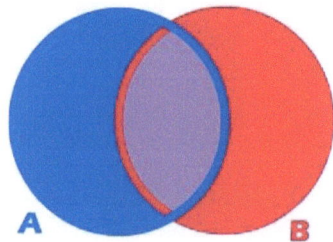

In contrast, the result of the Union of A and B

    A ∪ B

is shown in orange below.

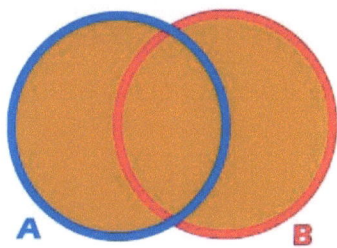

After removing the intersection (purple) from the union (orange), the following remains:

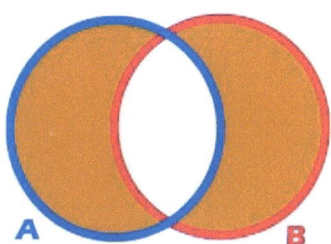

This is the visualization of the **XOR** operation when used on two sets. It was accomplished by using 3 other basic operators: union, intersection, and subtraction.

XOR-ing Three Sets May not be What You Expect

The wording "XOR-ing" is Mathematician slang for "applying the XOR operator".

Test your understanding by tracking the **XOR** of three groups: A, B, and C. Make sure that not all intersections are removed at the same time.

The notation for a XOR of three sets is A $\oplus$ B $\oplus$ C.

The calculation must start by choosing one pair first. It can be any of:

A $\oplus$ B, or

B $\oplus$ C, or

A $\oplus$ C.

Only after completing the XOR of one pair should the third XOR be attempted.

In other words, even though the order of operations can be random, as shown by the possibilities for parenthesis below, consistency is still required.

(A ⊕ B) ⊕ C

or

A ⊕ (B ⊕ C)

or

(A ⊕ C) ⊕ B

In all three cases the result will be the same.

Not resolving two sets at a time will yield incorrect results, let's see what can happen to clarify further.

Consider the three circles to be of colors: blue, red, and black. Where colors intersect, the shade is different or colors are blended. For example, a darker blue shows the intersection of the blue circle with the black.

The outline below depicts the union of all three circles.

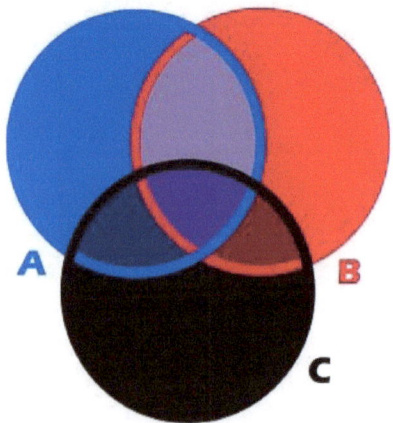

All possible intersections between the circles in purple, blue and red are shown below.

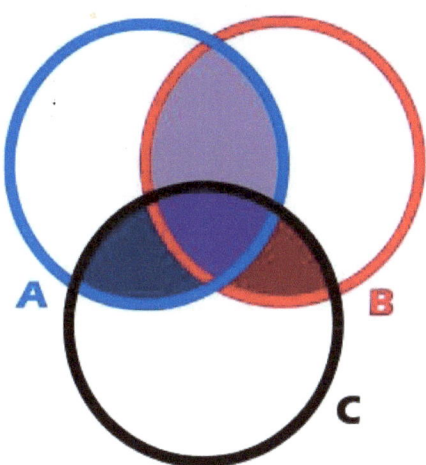

Removing the three intersections from the union together at once, results in:

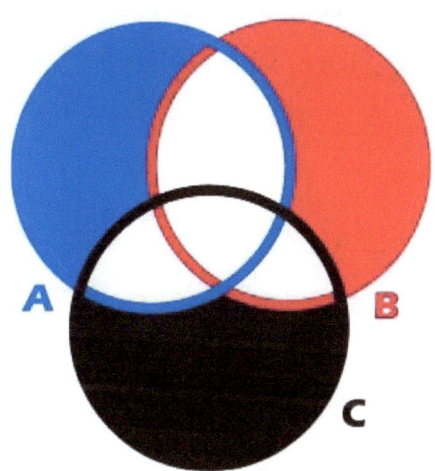

This is Incorrect!

This is **incorrect** because it does not depict the results of the XOR A $\oplus$ B $\oplus$ C. The correct solution is to XOR two circles at a time, followed by XOR-ing the third set with the result. For example, when starting with A $\oplus$ B, the outcome for the first pair of sets is.

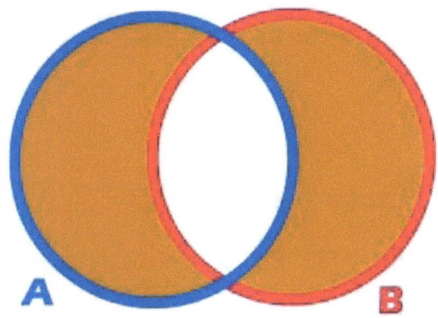

Adding the black circle C at this point can only intersect with the parts that were not removed so far. The yellow arrow shows the removed area which can no longer participates in intersections because it is simply not there.

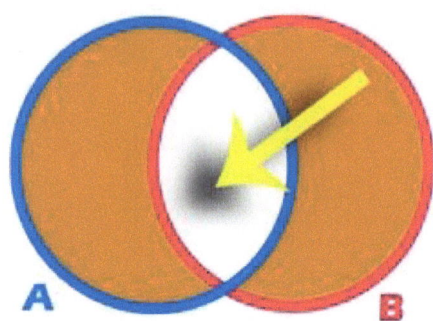

Colors were added again below using the same idea: intersections with the black circle make a darker shade. These intersections will be the pieces to be removed further.

Since A ⊕ B was calculated first, surrounding the expression with brackets is just a way to hint to that.

$$(A \oplus B) \oplus C$$

Emphasizing the order in which the operations were performed is a friendly notation though not necessary.

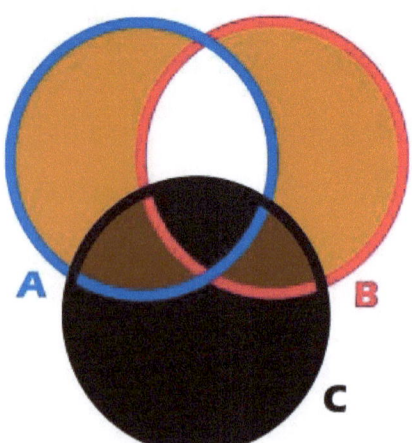

For the final result, the entire orange area is intersected with circle C. After removing the intersecting parts, the following structure remains:

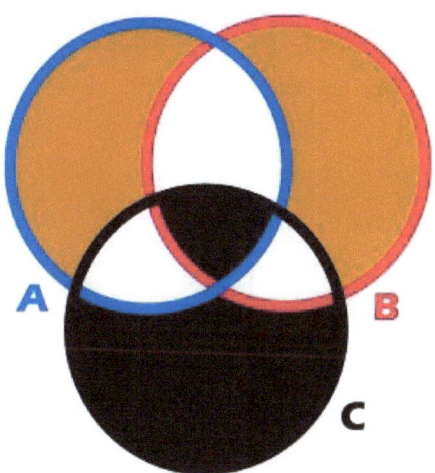

Let's compare the results of the two methods side by side, and see the difference clearly. On the left is the XOR when applied one at a time, on the right is the attempt to remove all intersections at once.

The correct version maintains a patch in the middle, the incorrect version doesn't.

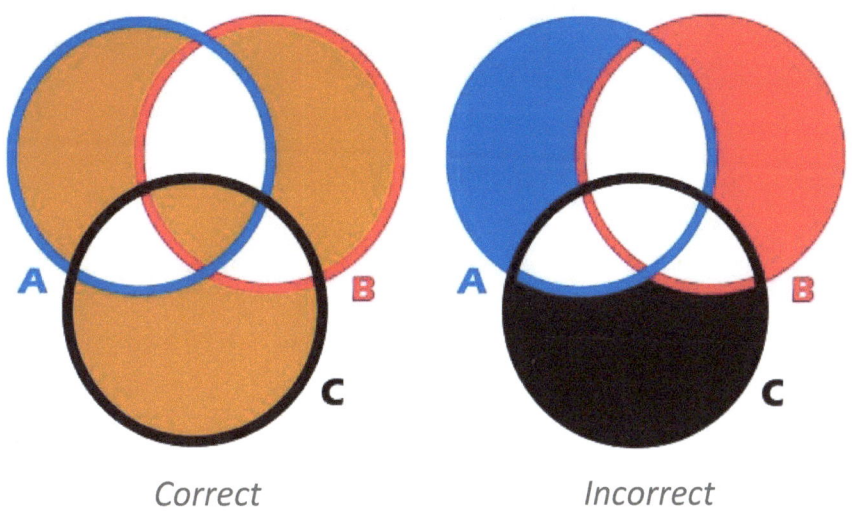

Correct *Incorrect*

Uses of Set Theory

Some interesting uses, or applications, of set theory are common in software engineering. For example, a frequent task is the organizing of various sets of countable objects.

Countable is emphasized as these are groups that are finite, contrary to groups like the set of integer numbers, which is infinite.

Managing groups in coding for software is easier when using ideas from Set Theory. All programming languages have grammar, otherwise known as **syntax**, that support basic set theory operations like: unions, intersections, merging (adding), subtracting, calculating complements, and more complex operators like **XOR**.

Set Theory is often used as an introduction to Logic.

Integrated Circuits on your computer boards contain physical elements that work together based on Set Theory rules. These rules put into action translations of analog signals into digital ones.

Take a Quiz on the Set Theory Chapter

Test your understanding of this chapter with a self-grading online quiz. Click the "View your score" link immediately after submitting your answers.

> Your response has been recorded.
>
> View your score
> Submit another response

Quiz submissions are anonymous. There is no limit to the number of times you can take the quiz.

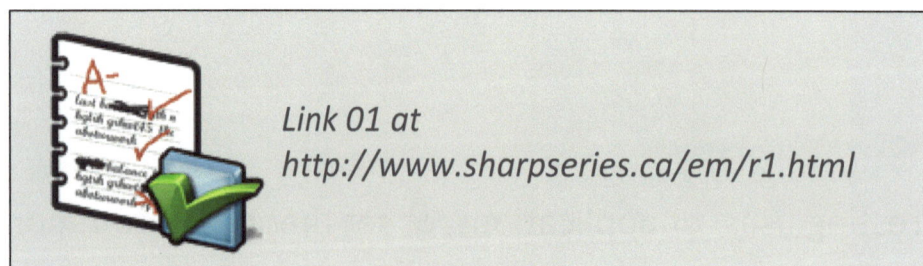

Link 01 at http://www.sharpseries.ca/em/r1.html

 Credit: Khan Academy Lecture Video	***Additional Resources, Explore Set Theory Further*** *Intersection and union of sets* *Link 02 at http://www.sharpseries.ca/em/r1.html* *(unaffiliated resource)*
Credit: University of Pittsburgh, Venn Diagram applet	*Explore logic operations on Venn Diagrams* *Link 03 at http://www.sharpseries.ca/em/r1.html* *(unaffiliated resource)*

Chapter 2: Fractals

A fractal is a geometrical shape ([2D or 3D](#)) that can be zoomed into or out of [infinitely](#) without losing the quality of the detail in its image. Fractals have patterns that repeat at infinite depths. At each depth, the fractal continues to be recognizable.

Let's compare a fractal image to that of a non-fractal before and after magnification.

Fractal:

Original *Zoomed in*

Non-fractal:

Original *Zoomed in*

In the upper row, zooming in shows the same detail on both left and right, it is difficult to tell the two images apart. The structure shows an infinite number of squares that turn slightly clock-wise with each step.

The results look the same at all depths.

In the second row, we are zooming in on Saturn. The inner elements of the image and their patterns do not repeat. Their shapes have no regularly repeating patterns regardless of the depth you are looking from. Saturn's image above is not a fractal.

Types of Fractals

A fractal is a mathematical idea. It is not real but rather imagined. In other words, it is said to be **abstract**.

The most accurate way to create fractals is by using computers, as they can draw patterns quickly and repeatedly, thus mimicking the infinite idea. Nonetheless even computer-generated fractals are [finite](), but this is as close as we can get to this **abstract** mathematical idea. For the fractal to be real the computer would have to continue drawing it forever.

Nature has lots of fractal-like patterns but none of them are true fractals due to the lack of [infinity]() in the pattern replication.

Patterns in nature repeat regularly for a limited number of times, some more than others, but it is still worth investigating their similarity to actual fractals.

Now that we have clarified the difference between the two, will continue to use the term **fractal** in both cases.

There are several types of fractals and can be sorted by the mathematics or by the computer methods used to draw them. This can get quite complicated and therefore is not included in this volume.

One option is to classify fractals based on the way they look. Some are noted for their identical elements and specific patterns that continuously repeat. Such patters are called [self-similarity](), and is explored in the section below.

Self-Similar Fractals

The way to discover how fractals behave is to use magnification to examine how the smaller parts look like. If the parts are identical or similar to the larger object at every magnification level, then the fractal is said to have **self-similarity**, or alternately, it is a **self-similar fractal**. All the parts assemble according to the same method.

Many self-similar fractals are found in nature. One example is the Romanesco Broccoli. This vegetable is not only self-similar, but is also known to be a **Fractal Pyramid** because its every element is a miniature identical, or rather very similar broccoli pyramid that is also made of smaller broccoli pyramids, and so on repeatedly.

An area in the image below (left) was selected with an arrow. Its magnified version is shown on the right.

Original   *Zoomed in*

The elements of the broccoli are smaller pyramids. These are made of even smaller broccoli pyramids. The parts look near-identical to the whole.

Let's look at other regular geometric self-similar shapes.

The image below is a computer-generated Fractal Pyramid called the **Sierpinski Pyramid**. It has self-similarity. Compare it to the broccoli above. Each element of the pyramid is made of smaller identical pyramids.

Other examples of self-similar fractals in nature are ferns and the branches of trees. Ferns have leaves that fit well in a triangular shape that repeats, and each leaf is made of many smaller versions of itself.

The branching of trees (below) is similar to the way smaller branches diverge. The smaller branches look like scaled-down versions of their trees. In this example, a part of the tree was carved out and turned clockwise so that it is positioned more like a tree than a branch. It would be difficult to know the image on the right is not a whole tree but rather a smaller branch.

The veins of many leaves follow a fractal pattern as well. Larger veins branch into smaller versions of themselves.

Fractals can be beautiful, and they are often used as elements in Computer Art like the one shown below.

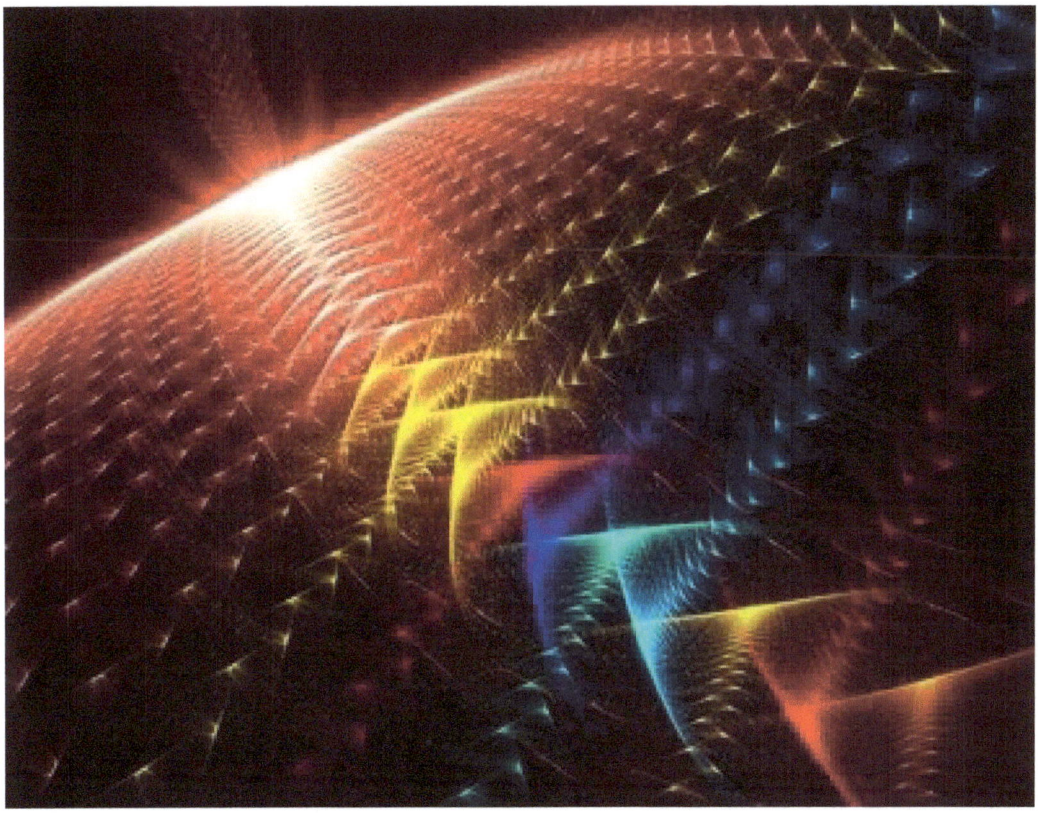

A few popular self-similar fractals are shown: The **Mandelbrot Set** and **The Koch Curve**. Both can be zoomed into indefinitely and will always look the same regardless of the depth of magnification. Such fractals are also called **Infinite Sets**.

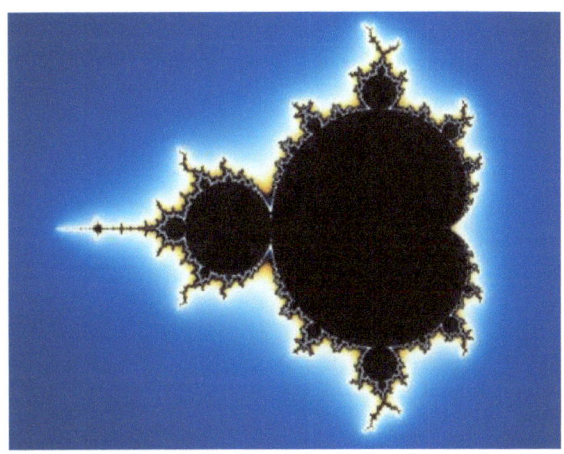

Mandelbrot Set   *The Koch Curve*

A famous computer generated self-similar fractal is the **Menger Sponge**. A flat, 2D version of the Menger Cube/Sponge is shown below as **Menger Squares**. These contain repeating perforations on increasingly smaller square subparts.

The 3D Menger Sponge design is fascinating for additional reasons and is discussed later in this chapter.

Non-Self Similar Fractals

Non-self-similar fractals have smaller elements that do not look anything like the original object. Pieces repeat infinitely in their mathematical version, or repeat for a limited number of times in computer generated fractals and those found in nature.

Below are a few examples of natural fractal patterns of **Phyllotaxy**. **Phyllotaxy** is the term for the way in which leaves arrange on the plant stem.

The artichoke

The leaves of the artichoke

The leaves of the artichoke (right) are not identical copies of the larger object (left). Leaf placement on the stem - the phyllotaxy - follows a steady predictable pattern. However, looking at the leaves, you may not necessarily guess what the shape of the parent plant is. Therefore, this is a **non-self-similar fractal** pattern.

The Aloe Vera plant   *The leaf of the aloe plant*

Just like the artichoke, the leaf of the Aloe Vera plant is not a replica of the larger object. Its phyllotaxy is a non-self-similar fractal.

This plant has leaves that become increasingly larger as the design moves into the outer layers. There is a mathematical formula that can predict the size of the next leaf and its position in the arrangement.

Phyllotaxy often exhibits mathematical patterns that place the leaves on the stem in a [Golden Ratio](), discussed in a later chapter in this volume.

Coast lines (below) are also examples of fractals in nature. The irregularities along the coast are there for a very large number of magnifications, but they do not repeat identically at each level, so while they are natural fractals they are not of the self-similar type.

Take a Quiz on the Fractals Section

Test your understanding of this chapter with a self-grading online quiz. Click the "View your score" link immediately after submitting your answers.

Your response has been recorded.

View your score
Submit another response

Quiz submissions are anonymous. There is no limit to the number of times you can take the quiz.

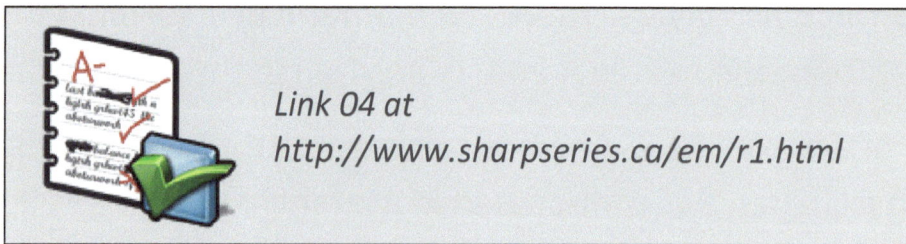

Link 04 at
http://www.sharpseries.ca/em/r1.html

The Menger Sponge

As mentioned earlier fractals are formed by fitting together geometric patterns that repeat but assemble based on specific methods. Such regular, repeatable methods are known as Algorithms. Formulas in general can also be seen as algorithms.

The following algorithm builds a **Menger Sponge (or Cube)**, a self-similar 3D fractal cube.

A few unusual and unique properties emerge from the construction of the Menger Sponge:

(i) the resulting cube has a near zero volume, and
(ii) the resulting cube has infinite surface.

The process and results are explained as follows.

Step 1:

Start with an empty cube.

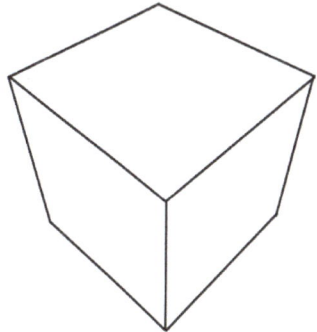

Step 2:

Divide the cube into 27 smaller cubes, by dividing each face of the cube into 9 smaller squares, as shown below.

If you were to add different color to each face, it would look a bit like a Rubik cube (below, right). Let's keep the color to help us track the changes.

Step 3:

Focus on the cells at the center of each face, as pointed by the blue arrows. Each has a smaller cube residing in that position.

Step 4:

Cut out, or perforate, the center cubes pointed by the arrows in the middle. The cube on the right shows the missing parts in white.

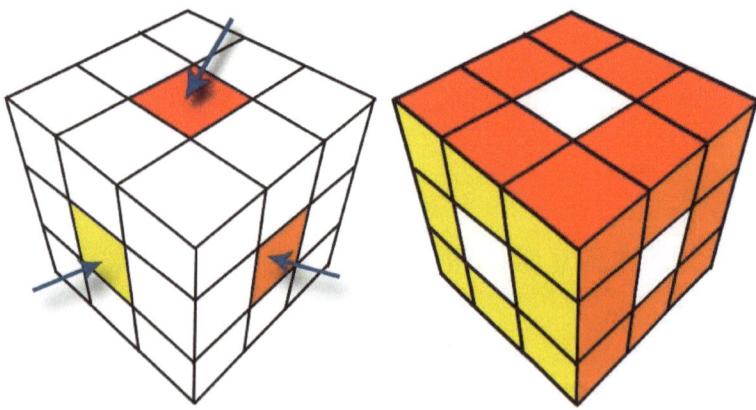

Step 5:

Pick one of the small colored cubes, for example the one pointed by arrow in the front top corner.

Step 6:

Repeat Step 2 on the selected cube.

Then follow the methods in steps 3 and 4 as well. A series of new perforations are made as shown below in white.

Repeat the process on every one of the remaining grey cubes. Do this on every side. The results are should look like this:

Step 7:

Using the current cube, return to Step 5 and repeat from there on an even smaller cube all the way to Step 6. Just like the previous time, pick the top front corner cube, where the arrow points below.

Repeat Step 7 as many times as you can, then move on to Step 8.

Step 8:

Decide when the cubes have become too small to continue the process, then stop.

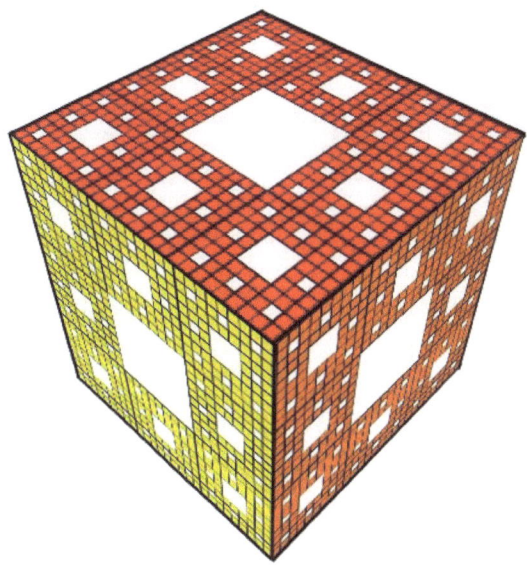

Conclusions

The following figure shows that many cubes of different sizes were removed throughout this <u>algorithm</u>, initially larger, then smaller, and we stopped when they were too small.

Each circled group shows one round of removals, the number of cubes affected, and their sizes.

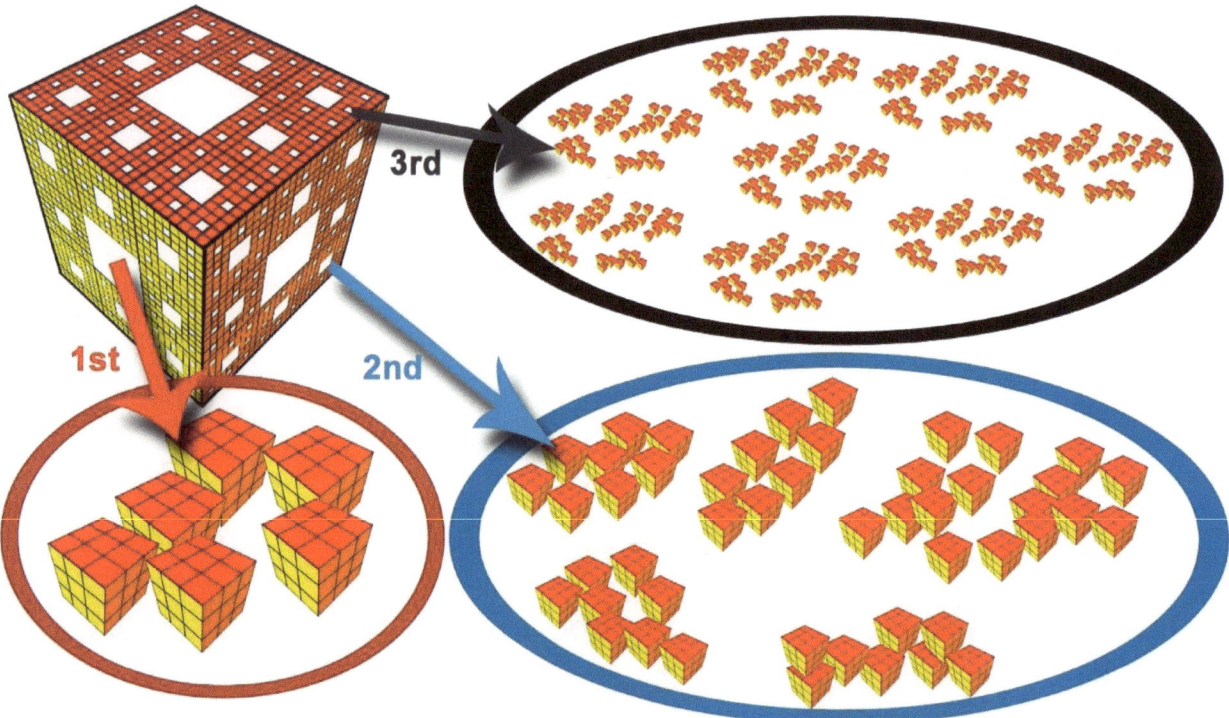

Notice the details in the figure above:

- Each round has its own circle, containing the number of extracted cubes in that phase.
- All cubes removed into one circle have the same size.
- The cubes in each round are smaller than in the round before (see the order of the rounds labeled on the arrows).
- The count of cubes in every round is much larger than in the round before.
- Each round accumulates larger volumes than the round before. The volume extracted by each round is increasingly larger.
- The original cube's surface area in each round is more perforated than the surface area of the previous round.

Next, a more precise approach is used to express these ideas in specific numbers summarized in the table below.

Round	Arrows show the removed cubes	Number of cubes removed in each repetition	Notes	The size of removed cubes in this round
1st	Red (bottom, left)	6x1 =6 cubes	1 cube removed from each face, times the 6 faces on the cube.	
2nd	Blue (bottom, right)	6x8 =48 cubes	8 cubes removed from each face, around the large gap in the middle, times 6 faces overall.	
3rd	Black (top, right)	6x8x8 =384 cubes	8 tiny cubes around every mid sized gap on each face. And there are 8 such around the larger gap in the middle.	

After removing every possible tiny cube repeatedly, the resulting object has an extremely large number of holes.

Imagine the piles of cubes removed from the original as calculated here. There is a pattern that starts to emerge in the second round:

1st round	6 cubes removed	1 x each face
2nd round	48 cubes removed	8 x the cubes removed in the previous step = 8 x 6
3rd round	384 cubes removed	8 x the cubes removed in the previous step = 8 x 48
4th round	3,072 cubes removed	8 x the cubes removed in the previous step = 8 x 384
...	...	**8 x the cubes removed in the previous step** = ...
8th round	12,582,912 cubes removed	
...	...	...
10th round	805,306,368 cubes removed	...
...	...	...

In a true fractal, the removal of cubes continues to happen infinitely. The piles of cubes removed in the end is so tremendously large that nearly equals the entire volume of the original cube. Every tiny cube encountered by the <u>algorithm</u> gets divided into smaller cubes and is perforated removing more of its contents.

With so much of its content removed, the original cube is left with nearly no volume at all, in other words a **near-zero volume**, and is now made of only a collection of thin sheets that crisscross everywhere in the inside, very similar to a real sponge.

Below is a computer-generated rendering of a Menger Sponge.

While it is easy to understand the near-zero volume of the Menger Sponge by simply looking at the enormous amounts of removed cubes, it is a bit more difficult to understand why its surface is near-infinite. How can you obtain infinite surface area from a finite volume?

To dive into this concept, let's examine what **Surface** is, and what happens to it during the removing of the smaller cubes.

Volume and Surface Area

All objects, except for points, whether 2D or 3D, have **Surface**. For 2D objects, the surface is the object itself.

For example, a sheet of extremely thin paper is very similar to a 2D object, but only if you think of it as having a single side, that is a *Front* but no *Back* as shown below. This may be initially counterintuitive, but once you consider a sheet of paper to have two sides it instantly becomes clear it is 3D, and there must be a space enclosed between the Front and the Back, no matter how thin.

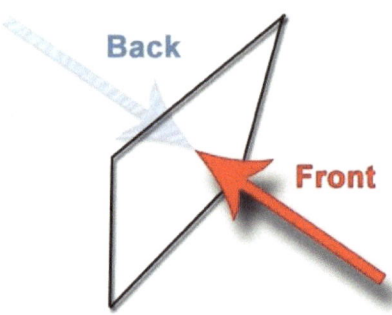

In Math, examples of true 2D elements are dots and lines. They are considered to be abstract.

In real life, nothing is quite 2D as all things have depth, which makes them 3D. It is a matter of inventing better devices, such as microscopes, to observe the actual 3D nature of all things.

Therefore, a 2D sheet has one surface, and it is single sided. A 3D sheet has multiple surfaces, such as front, back, width, and height.

For this section, it is of particular interest to investigate the surface area of 3D objects. We'll look at cubes in general first, to investigate the surface of the Menger Sponge.

The surface of an object is similar to the peel of a fruit, the skin of a balloon, the paint on a wall, etc. You can think of it as an imaginary, abstract layer that separates an object's inside from its exterior.

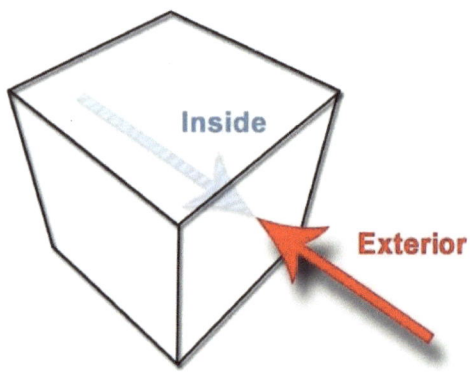

The more bloated an object, the more abundant its overall surface becomes. For example, balloons increase their surface area while they are being inflated. The interior space is growing as it is being inflated by the added air. At the same time the surface of the balloon stretches and becomes larger too.

Just like balloons, all 3D objects have an inner space that can be filled and measured. Imagine filling such an object with air like in the case of the balloon, water, miniature cubes, or beads. Water and sand can be weighed, cubes and beads can be counted. In each case, you can connect the magnitude of the inner space to numbers that can then be compared amongst objects. The larger the number, the more abundant the inner space is.

The inner space discussed here is called **Volume**. All 3D objects have volume. In contrast, 2D elements like lines and points, have no volume.

Surface Area of the Cube

The total surface area of the cube is the sum of all its surfaces.

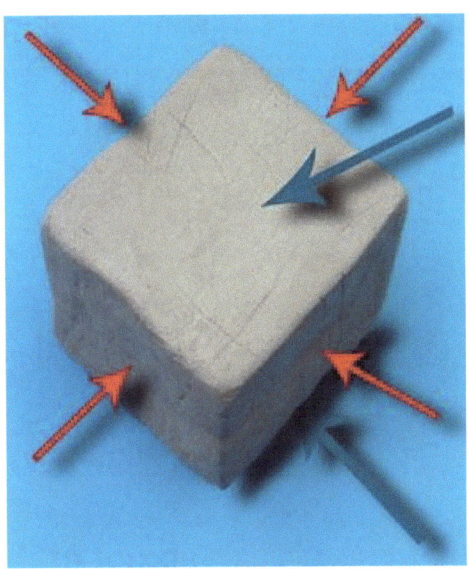

In the example above, 2 blue arrows point to the top and the bottom, and 4 red additional arrows point to the sides. Together this accounts for all the 6 faces.

All sides of the cube are identical squares. The area of each face is based on a formula that takes any of its edges squared (i.e. multiplied by itself).

Assume an edge was measured to be equal to "a" cm. The area of one face is then: **a x a cm² = a² cm²**.

The total surface of the cube equals the sum of all its squares, i.e. 6 times the surface area of one face, in other words that is **6 x a² cm²**.

Since all the Menger algorithm does is to remove cubes from the volume, the consequence is the removal and addition of assorted squares from faces and into the indentations formed. Having to only deal with squares will make tracking the changes to the overall surface area especially easy.

Increasing the Surface Area of the Cube

Let's examine how the surface area of the cube grows during the making of a Menger Sponge.

In the Menger algorithm, at every round a smaller cube is carved out of the middle of every side of various cubes. Such a cube becomes as shown here:

Examine the surface in the spot where a small cube is carved out. The pictures above show before and after states of the carving. If the edge of the removed cube equals "b" cm, then the surface area covering that spot equals to **b^2 cm²**. The cube lost this surface area in the middle and is now replaced by a hole.

The missing material reveals new inner surfaces. The arrows below point out the back, top, bottom, right and left of the cube-shaped gap. These are the walls of the carved opening. Each of these walls is square shaped, identical to the sides of the cube removed.

The count of all new surfaces in the indented space is five. Each image in the set that follows shows one of these newly revealed areas.

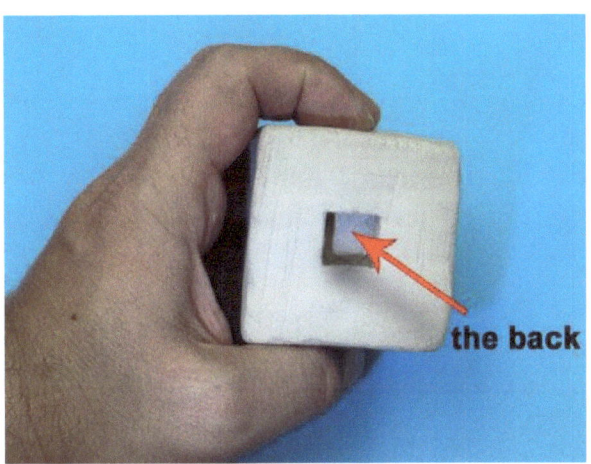

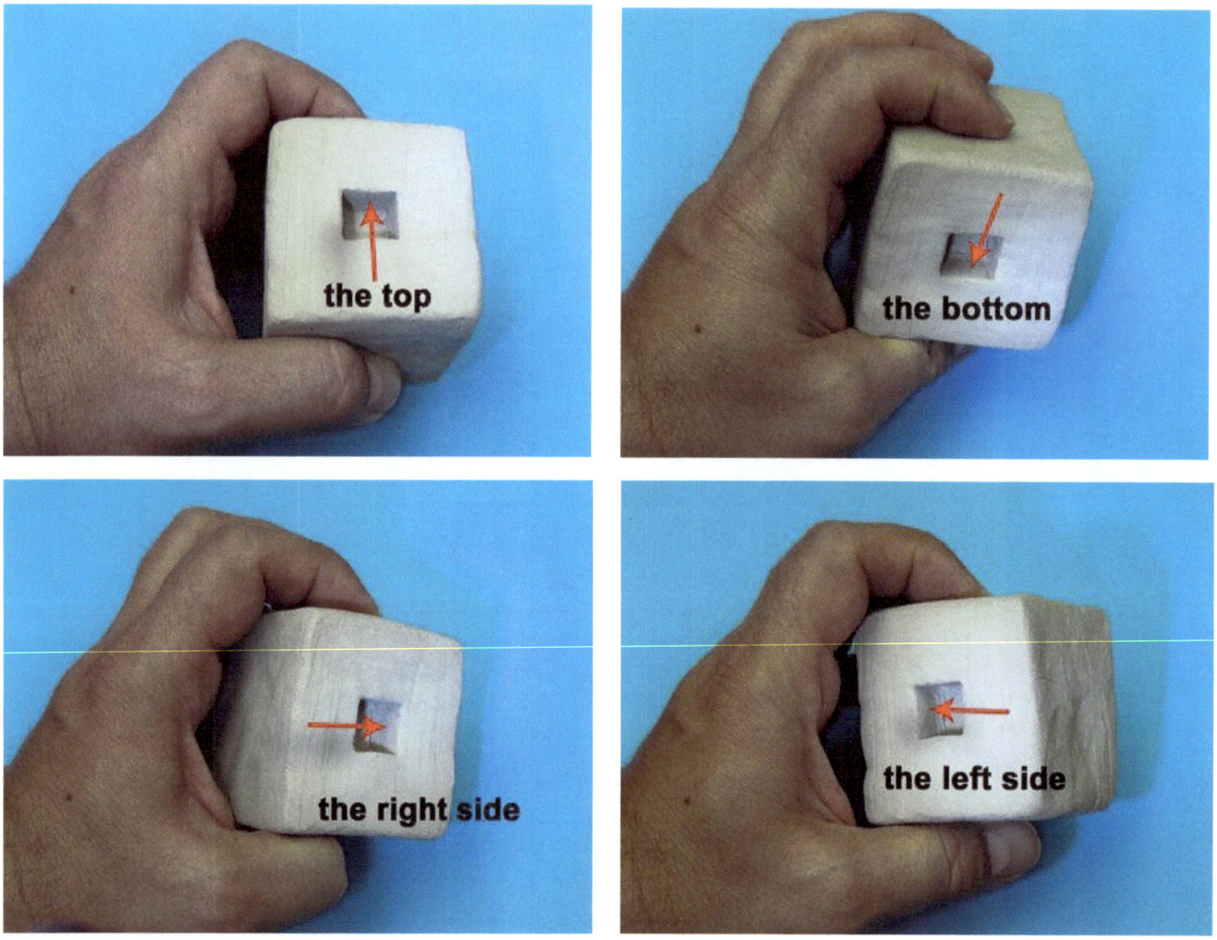

Let's summarize. The carving removes one surface belonging to the removed cube, but introduces 5 new ones by exposing the inner walls. The result is the same as overall **adding** 4 new surfaces (= 5 -1).

Therefore, during the Menger Sponge algorithm, every time a bit of the surface area is carved out, the cube's surface area grows by 4 times the surface removed.

We've seen in the pattern of cube removals earlier that with each round a larger number of cubes is detached. Multiply the number of cubes removed by 4 to get the new surface area they contribute.

1st round	6 cubes removed	6 x 4 = 24 new surfaces added.
2nd round	48 cubes removed	48 x 4 = 192 new surfaces added.
3rd round	384 cubes removed	384 x 4 = 1,536 new surfaces added.
4th round	3,072 cubes removed	3,072 x 4 = 12,288 new surfaces added.
...	...	**4 x the number of cubes removed in this step** = ...
8th round	12,582,912 cubes removed	12,582,912 x 4 = 50,331,648 new surfaces added.
...	...	...
10th round	805,306,368 cubes removed	805,306,368 x 4 = 3,221,225,472 new surfaces added.
...	...	...

Notice how much faster the added space grows with every round of removals, when compared to the number of cubes detached.

It becomes easier to see that with such a massive number of new surfaces added in every round, very quickly the Menger Sponge gains a tremendous amount of surface area. Continued for a large number of times, or rather infinitely, the never-ending adding to the overall surface area, makes it infinite as well.

If this still bothers you, try to remember that the cube already contained an infinite number of surfaces within itself. This is due to being able to split any sheet further into smaller, thinner panes infinitely. The Menger algorithm simply reveals them into the open.

The near zero volume of this cube and its near infinite surface are the very amazing hallmarks of the Menger Sponge.

Usefulness of the Menger Sponge Model

The Menger Sponge has impact on various sciences other than Math. For example, it is used in attempts to model extremely porous materials in physics and chemistry.

This structure is also useful in the synthesis of super water-repellent and highly oil-repellent fractal surfaces.

The Menger Sponge design could prove useful in future disciplines that require massive surface areas.

Take a Quiz on the Menger Sponge Section

Test your understanding of this chapter with a self-grading online quiz. Click the "View your score" link immediately after submitting your answers.

> Your response has been recorded.
>
> View your score
> Submit another response

Quiz submissions are anonymous. There is no limit to the number of times you can take the quiz.

Link 05 at http://www.sharpseries.ca/em/r1.html

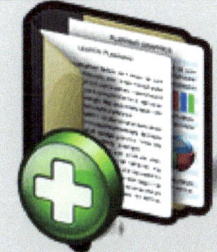

YouTube Video Credit: Massachusetts Institute of Technology (MIT)

YouTube Video Credit: Serveurperso

Additional Resources, Explore the Menger Sponge Further...

Mega Menger: Building a Menger Sponge at MIT

Link 06 at http://www.sharpseries.ca/em/r1.html (unaffiliated resource)

Trip inside a Menger Sponge level 14 (3D fractal)

Link 07 at http://www.sharpseries.ca/em/r1.html (unaffiliated resource)

The Sierpinski Triangle

A famous self-similar fractal is the **Sierpinski Triangle** in <u>2D</u>, and the **Sierpinski Tetrahedron** in <u>3D</u>.

Here is how to create the fractal for the triangle.

We'll use color to help visualize the difference between <u>consecutive</u> steps. However, the colors denote areas that are cut out of the triangle and is not part of the final design. At the end, all colored triangles get literally cut out from the original.

In this algorithm, new elements are added inside white areas and no additions are permitted into the already colored ones since those are deleted spaces.

Step 1:

A white <u>equilateral</u> triangle is given.

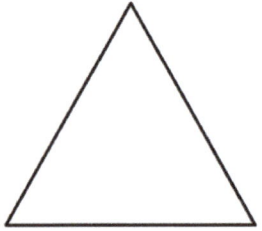

Step 2:

An <u>equilateral</u> red triangle is drawn upside down in the middle. The new triangle is placed with its base up and tip down. The base divides the side edges in two equal parts, each and the tip divides the bottom edge equally as well.

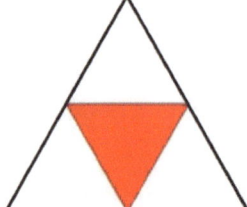 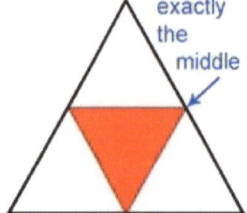 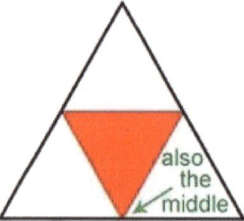

Three smaller white triangles have formed around the red one (blue arrows below, on the left).

The red triangle must be drawn correctly, precisely in the middle. Then all triangles, red and white are ensured to be equilateral and identical to one another.

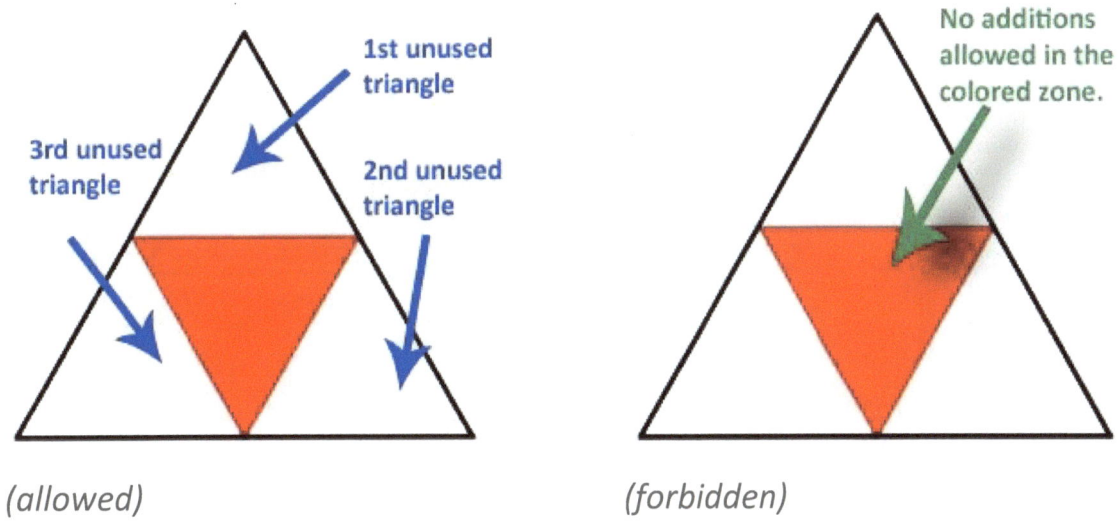

(allowed)   (forbidden)

Step 3:

The previous step resulted in a new collection of triangles of two types: colored and white.

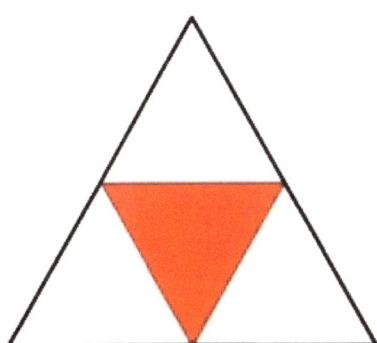

Go back to Step 1 and repeat on every white triangle found in the current step. Return here when done.

The expectation is that Step 1 added 3 new upside-down triangles. These will be equilateral and colored. Orange was used instead of red. Orange differentiates triangles of this step from those made by the previous step. Three equal white triangles surround the new orange triangles, shown by the green arrows on the right below.

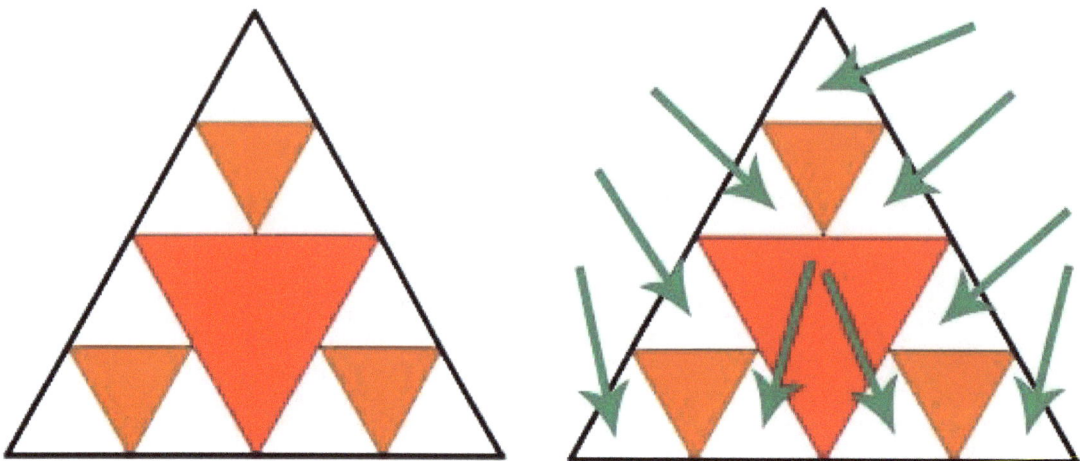

Step 4:

Go back to Step 1 and repeat on every white triangle found in the current arrangement. Return here when done.

The conclusion is that in this step as well, each white triangle (see green arrow)

is replaced by 1 colored triangle and 3 white ones around it (blue arrows).

The number of white triangles in the previous step is multiplied by 3. For example, if you start out with 9 white triangles (left) this step yields 9 x 3 = 27 white triangles (right).

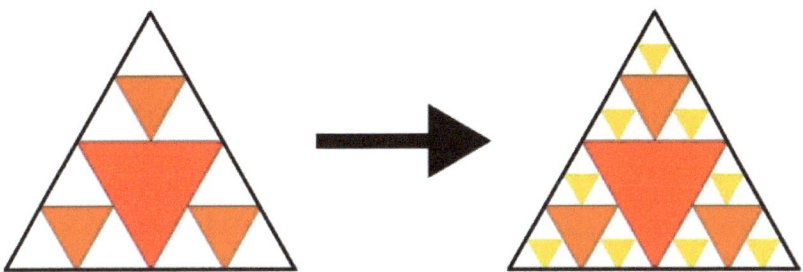

Step 5:

Go back to Step 1 and repeat on every white triangle found in the current arrangement. Return here when done. The result is shown below.

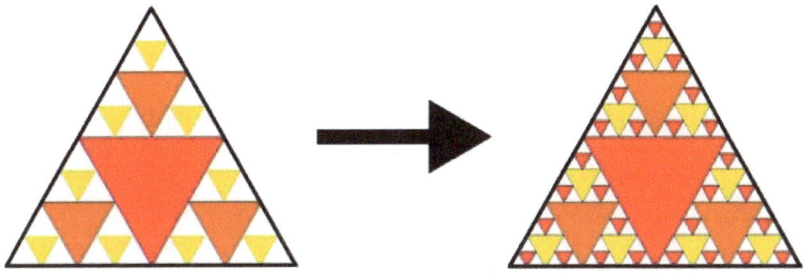

Step 6:

Go back to Step 1 and repeat on every white triangle found in the current arrangement. Return here when done. The result is shown below.

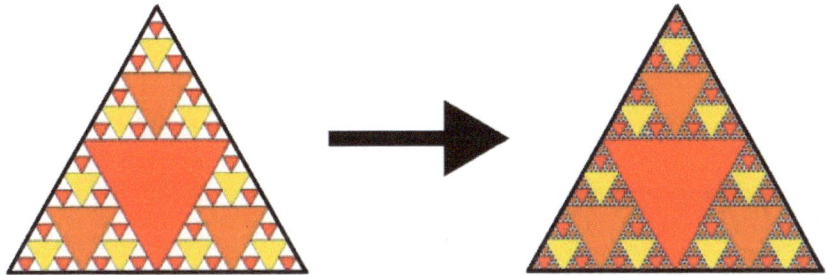

Decide when to stop. If newly drawn triangles are small as dots, there is no point to continue the process.

Remove all colors used earlier, which were only there to help separate the work done by consecutive steps.

What is left in black-and-white is the classic look of the Sierpinski triangle. The black parts represent what is left of the area of the triangle, and the white parts in this case show where pieces were cut out, i.e. the fabric of the triangle, for example paper or other materials, is missing.

Imagine this process to continue infinitely. With computer-generated fractals, it is easy to zoom in further into a microscopic world. For the Sierpinski Triangle, the zoomed image looks always identical to the larger object, which is why this fractal is self-similar.

Area and Perimeter of the Sierpinski Triangle

When examining the progressions of the Sierpinski Triangle it is evident that increasingly more area of the triangle is removed in every step.

When the number of steps is large, or infinite in theory, the area has so many holes that it becomes near-zero. While the area decreases, the perimeter of the triangle increases in tandem. Let's track how this happens from step to step. The following triangle,

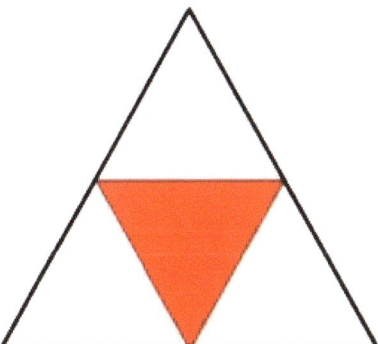

has two types of perimeter lines: (1) lines surrounding the exterior of the triangle, and (2) lines tracing the inner edges around the red region.

Since the red triangle is actually a hole from a cutout, the lines tracing it are edges and part of the perimeter. The collection of all edges belonging to the perimeter is shown in blue.

The next triangle in the sequence has additional perforations in orange. Around them new edges form. The arrangement looks so:

New edges are added to the existing sum of all edges that make the total perimeter.

The blue colored edges show the original exterior perimeter plus its newly added edges tracing the colored perforations:

To differentiate between the perimeter of the earlier step and the current additions, purple color was for the earlier ones, and new edges are shown in blue.

The blue lines are what give the increase in the perimeter length. Each step in the algorithm provides additional cuts that open new edges, so the number of edges is always on the rise causing the perimeter to increase. With infinite steps the total number of edges and the perimeter both become infinite.

The end result of the Sierpinski triangle algorithm is thus a near-zero surface area with an infinite perimeter, and an infinite number of edges and vertices.

The Sierpinski Tetrahedron

When building a 3D version of the triangle the result is a Sierpinski Pyramid (or Tetrahedron). A Tetrahedron is a triangular pyramid, both wordings will appear in this book and are interchangeable in this example.

This pyramid has a few similarities with the Menger Sponge and some significant differences as well. Their algorithms progress similarly, one removes cubes and the other cuts all tips halfway through the pyramids from a previous step and progress so infinitely.

The **Sierpinski Tetrahedron** algorithm is described as follows.

Start with a simple tetrahedron. The two images show the same pyramid, from two angles: back and front.

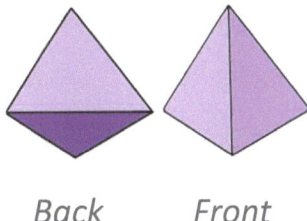

Back   Front

Step 1:

Choose a pyramid. In the first step there will only be the one, and more will be available in later steps. Find the middle of every edge, draw lines tracing the middles as shown, and prepare to scissor the tips off along the dotted lines.

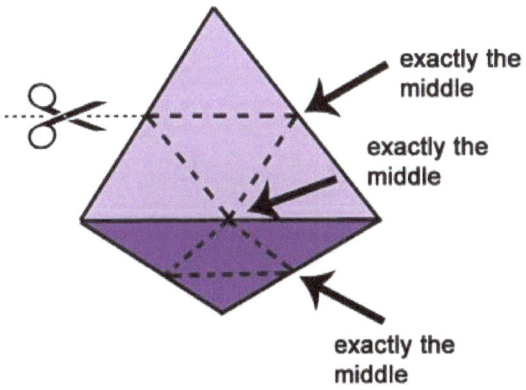

65

Before you cut the tips off, color them differently and count how many tips are accounted for. Here is the coloring for this example: red on top, green for the left tip, blue on the right and yellow at the bottom in the back. This counts 4 tips in total.

The unchanged space that is still purple, is not part of any tip and thus was not re-colored. The purple parts are to be removed.

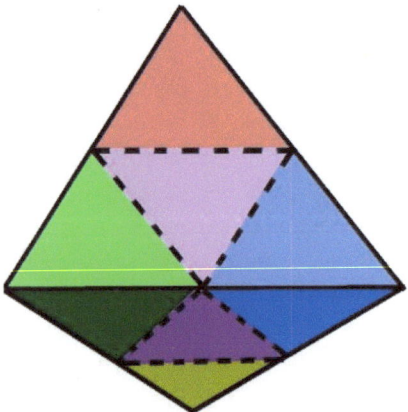

After completing the cuts, the pyramid is left with an arrangement of the colored tips, and the earlier purple space in between is abandoned.

To see the shapes from two different angles a flower was placed between the yellow and blue tips. This should help visualize the turning around of the arrangement from back to front. The group was then turned to face the reader, facing the flower and the two faces it overlays on.

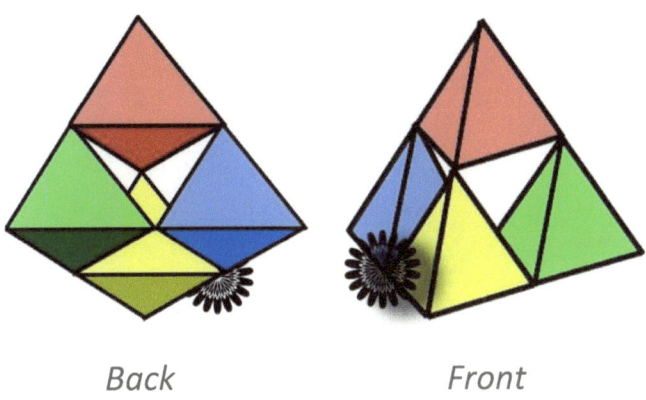

Back   Front

Step 2:

Bo back to Step 1 and repeat on every colored tetrahedron found in the current shape. The result is further sliced into smaller pyramids by the same method.

Return here when done with Step 1.

Before Step 1 is started   After Step 1 is completed

Step 3:

Bo back to Step 1 and repeat on every colored tetrahedron found in the current shape. The result is further sliced into smaller pyramids by the same method.

Return here when done with Step 1.

Before Step 1 is started   *After Step 1 is completed*

Step 4:

Repeat the previous step any number of times, then you stop.

Surface Area and Volume of the Sierpinski Tetrahedron

Earlier it was noted that the Menger sponge has a near-zero volume, and an infinite surface area. A comparison with the Sierpinski Tetrahedron follows.

The Volume

Assuming the algorithm goes on forever, in every step, pyramids are converted into smaller versions of themselves by separating the middle sections from the tips. Thus, more empty space is created in between. The arrows below show a few of these empty spaces.

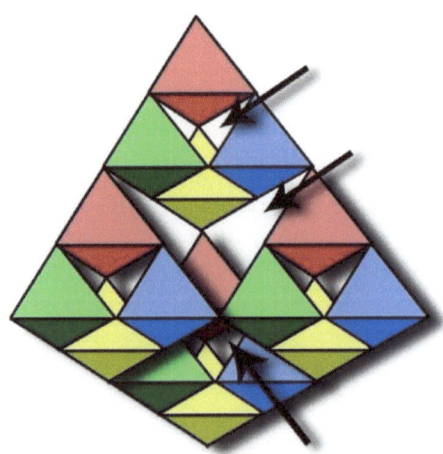

Eventually the volume of the resulting pyramid becomes microscopic. The outline of the outer tetrahedron will be filled with pyramids the size of dots or smaller, the rest will be mostly empty space. Thus, **the resulting volume of the overall shape is near-zero**.

The Surface Area

The Menger Sponge has an infinite surface area, intuition may dictate that the Sierpinski Tetrahedron would be such too. This is not the case.

To figure out how changes in surface area occur, let's see how they happen between two consecutive steps.

An easy step to examine is transitioning from 1 tetrahedron to 4. That is the first step of the algorithm.

The conversion is shown below.

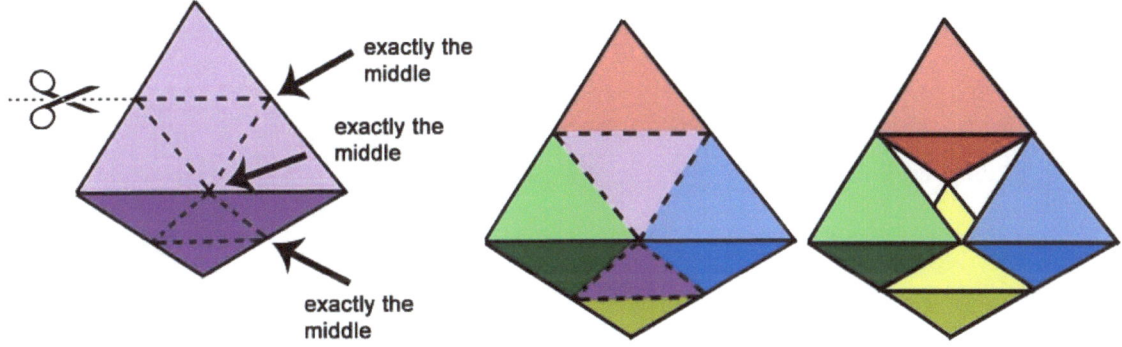

The important questions are:

1. How much surface area was lost in this transition?
2. How much surface was gained?
3. What is the equivalent change in total, otherwise known as **Net Change**.

Using color to track the changes makes the process easier. The volumes lost are purple. They reside in the inner space that did not belong to any of the tips. Purple tetrahedrons (above middle) turn into empty spaces as they get discarded. In terms of lost surface area, the number of purple triangles deleted from the surface is 1 for each face of the original tetrahedron. **There are 4 faces, so the total purple areas lost are also 4**.

Sizes of the purple, red, green, blue, and yellow triangles are identical. These are all equilateral triangles, identical sized edges, same angles, and thus identical surface areas.

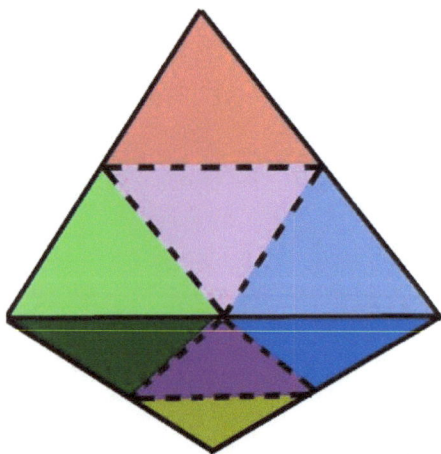

The tetrahedron didn't just loose surface, it also gained it by exposing more of the inner space.

The number of faces gained so is one inner face for every colored tip. Let's clarify.

The arrows below point at areas that were there before and after the carving, so we won't count these as new.

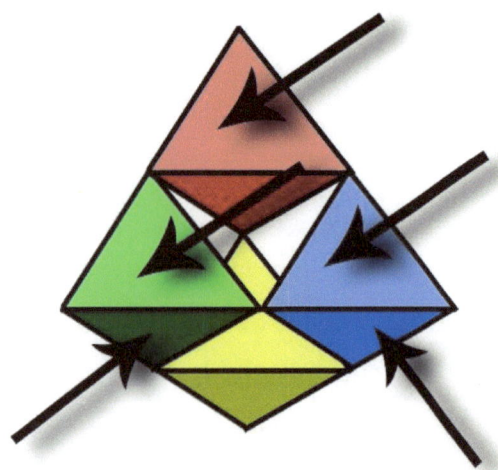

On the other hand, the bottom of the red pyramid (arrow 1 below), and the inner face of all other 3 pyramids: green, blue, and yellow, were not in the open in the previous structure, they're all new surface areas. Red and yellow are easy to count in the image on left. Let's turn the tetrahedron all the way around (below right) to count the new blue and green faces as well. Green and blue are counted in the image on the right after the pyramid was turned all the way. The flower is only there to help visualize the 3D objects being turned to face the reader.

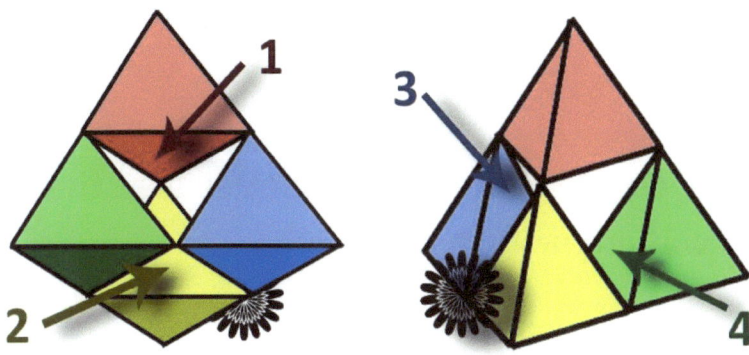

In conclusion **4 new inner surface areas were gained as of the equilateral triangles.**

Let's review what the total, or rather net change is: 4 purple triangles were lost, 4 inner colored triangles were gained, all these triangles were part of the same step. The surface area of both lost and gained equilateral triangles were equal to each other. Then with 4 lost and 4 gained, there is no net change at all.

New Area = Previous Area + 4 triangles − 4 triangles = Previous Area + 0 triangles

Therefore, the area of the new object is exactly the same as the area of the previous one. The calculation repeats the same way between any consecutive steps. The result will always be: the number of triangles added and removed from the overall surface area is equal. This means the surface area remains the same. The proper mathematical wording for this is: the surface remains **constant** throughout the steps of the algorithm.

Conclusion

A Sierpinski Tetrahedron at the end of its algorithm has the following features:

- Infinite number of components, i.e. smaller tetrahedrons, edges, and vertices.
- Infinite perimeter, equal to the sum of the lengths of an infinite number of edges.
- Near-zero volume.
- Constant surface area, equal to the original surface area the tetrahedron started with in the first step.

Take a Quiz on the Two Sierpinski Algorithms

Test your understanding of this chapter with a self-grading online quiz. Click the "View your score" link immediately after submitting your answers.

Your response has been recorded.

View your score
Submit another response

Quiz submissions are anonymous. There is no limit to the number of times you can take the quiz.

Link 08 at http://www.sharpseries.ca/em/r1.html

YouTube Video Credit: Vi Hart

Additional Resources, Explore the Menger Sponge Further...

Doodling in Math Class: Binary Trees

Link 09 at http://www.sharpseries.ca/em/r1.html (unaffiliated resource

Make Your Own Self-Similar Fractal, Create an Algorithm

Create your own fractal pattern by inventing and following a few rules.

You can use pen and paper or any drawing software in which copy and paste can be used to repeat the elements you are working with.

Compare your work to how the Sierpinski triangle was made above. Think about how many of the steps repeat, and keep jumping, or looping back to the same step. Make sure your method has an idea that repeats and demonstrate the steps in a diagram like the one shown in the example below.

Multiply x 2	Place on top		Assemble the pieces

The resulting geometrical shapes should show either a growing number of elements, shrinking spaces, or both.

When steps repeat so orderly, we call the method an algorithm, just like we did in the case of the Menger Sponge and the Sierpinski Triangle.

Debate what rules were applied in the steps of this algorithm.

The end result is a non-self-similar fractal pattern since the individual leaves do not look like the larger object.

Examining Algorithms Further...

Think of an algorithm to be a formula for anything that can be accomplished in steps. The steps are composed in a way to help someone else get the same results as you.

Cooking recipes are good examples for algorithms. Their purpose is to allow cooks to bake foods that taste the same every time.

In Computer Science, Algorithms are the most important and interesting elements. Algorithms restrict themselves to specific parts of a program and express the logic that solves a problem.

They often require numerous repetitions, like in the case of generating Fractals. This is the reason computers are used to draw most fractals rather than humans, as the job is so tedious and repetitive.

For any given problem, there can be many algorithms to solve it. These can differ based on the priorities you want to apply:

- solving the problem in the fastest possible way,
- having the smallest number of errors,
- using the smallest amount of memory.

Several algorithms have become famous due to their high speed and correctness and extreme difficulty to improve upon.

Examples of famous programming algorithms are: Quicksort for sorting numbers, Dijkstra's Algorithm for finding the shortest path in a network, and the RSA encryption algorithm, and there are many more.

In Math two of the most famous algorithms are: the long division algorithm, the greatest common divisor (GCD) of two numbers known also as Euclid's algorithm, and divisibility questions for some primes.

Algorithms are typically written in programming languages such as C++, C#, PHP, Java, etc., but can also be represented graphically, see below.

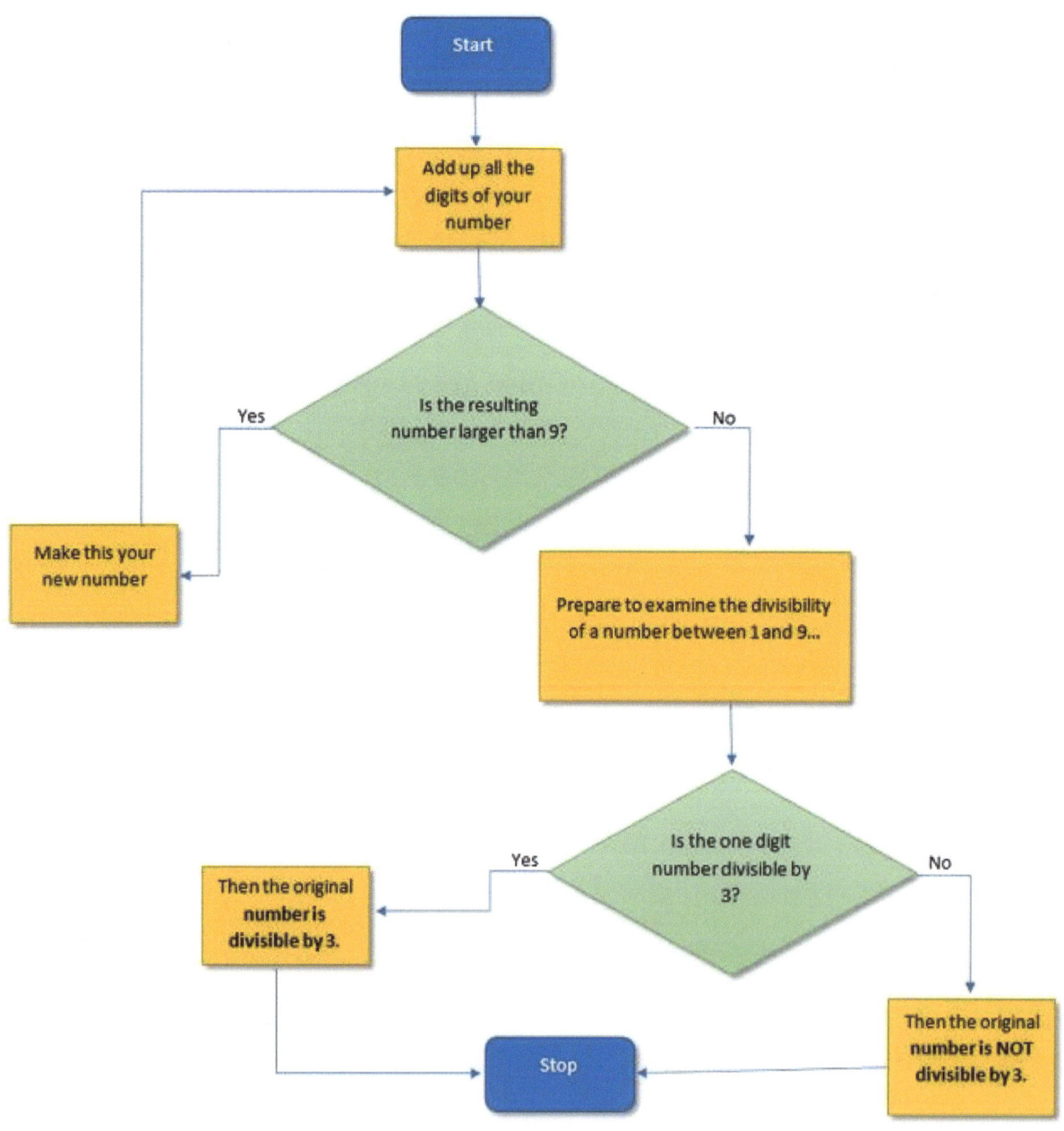

Such graphical representations of algorithms are called **Flow Charts**. Flow Charts are expressed through a series of shapes connected through arrows. The Flow Chart above shows a simple algorithm for determining divisibility by 3.

The arrows show how decisions are made from one shape to the other. The shapes must be used according to agreed-upon conventions: ovals are used for the start and end of the method, rectangles state one basic step such as count, add, subtract, and diamonds are used to ask simple Yes or No questions.

Flow Charts are used to emphasize the logic of an algorithm when the choosing of a particular programming language is not important.

The Flow Chart on the next page is an example for the algorithm that checks if a number is divisible by 3. Follow the logic and flow, test a few numbers for example: 123, 4212, 7980, and 71.

You can create your own flow charts using any of several free Flow Chart Tools online, one such example is "Google Drawing" at: https://drawings.google.com, all you need is a Google account/email.

Use a search engine to find other free online Flow Charting tools and experiment with creating diagrams for instance try describing the method you use when looking up a word in the dictionary.

Watch these videos that explore algorithms further.

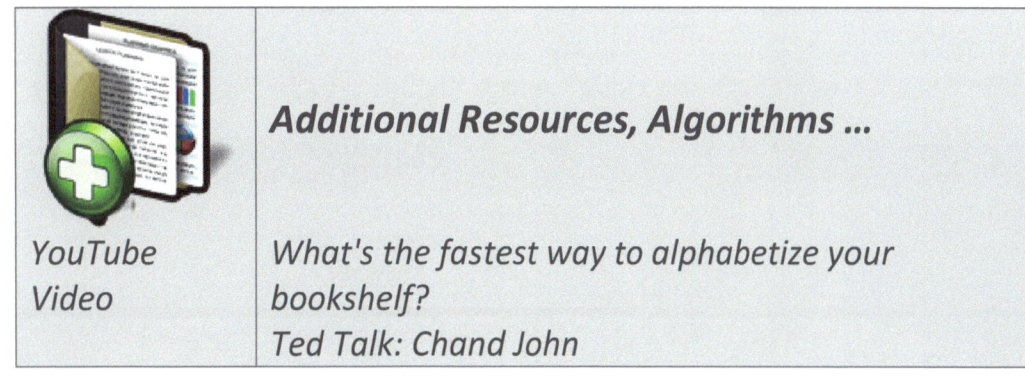

YouTube Video

Additional Resources, Algorithms ...

What's the fastest way to alphabetize your bookshelf?
Ted Talk: Chand John

Khan Academy Lecture, Video	*Link 10 at http://www.sharpseries.ca/em/r1.html (unaffiliated resource)* *What is an algorithm and why should you care?* *Link 11 at http://www.sharpseries.ca/em/r1.html (unaffiliated resource)*

Chapter 3: The Fibonacci Sequence and the Golden Ratio

A [Ratio](#) is the result of a division. A ratio can also be said to be a specific [proportion](#).

The Fibonacci sequence is a series of numbers in a specific pattern. You can predict new numbers in the series based on the two previous ones by adding them up. To visualize how this works, imagine numbers are placed in colored squares. Take two numbers, in the blue and orange boxes first. Together they start forming a sequence.

The next number in the sequence is calculated by adding the number in the blue and orange boxes. Let's say the result is placed in the new pink square.

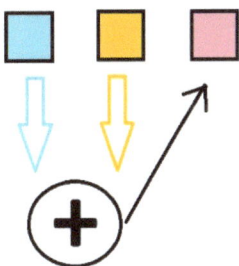

Pretend the next number in the sequence gets the green square. It is obtained by the same rule, add the last two elements together. In this example it is the numbers in the orange and the pink squares.

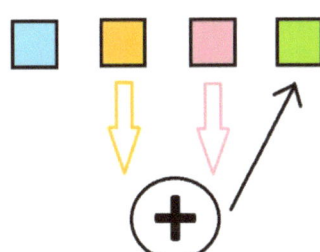

Examine the sequence with actual numbers. Let's say the blue and pink squares both contain the number 1. The sequence evolves as follows:

1, 1

1, 1, 2

1, 1, 2, 3

1, 1, 2, 3, 5

...

The Fibonacci numbers are getting larger with every calculation. This growth is similar, though not identical to another rhythm known in Math since ancient times as the Golden Ratio.

The Golden Ratio is the division between an element and the previous number in the series. The division of consecutive pairs for Golden Ratios is precise, always equals the same number: **1.6180339887…** One can truly think of it as "golden", in other words unchanging.

In contrast, Fibonacci numbers divide their pairs of consecutive elements to results that are only similar to the Golden Ratio, but not quite as precise. The Fibonacci ratios differ in tiny bits from one another, wobbling very closely around the Golden value, but are still considered to be only close approximations.

Due to its high precision, the Golden Ratio is not likely to be found in nature. However, its closest approximation, the Fibonacci Ratio, is.

Several natural fractals arrange themselves apart based on Fibonacci rules. For example, the way leaves arrange on many a plant stems (Phyllotaxy), the ratios in which some chemical compounds include each other, the shapes of some crystals.

Fibonacci ratios are often found in art, music, product design and architecture, because the results are thought to be esthetically pleasing.

The next example shows a **Fibonacci** arrangement and uses leaves to show the distance between them. The method described is an algorithm.

Step 1:

Draw a small square. Draw a black line along the edge on the top.

Add a leaf to the top left of the black line.

Step 2:

Start at the corner where the previous leaf was placed. Draw a red line perpendicular to the black line in a counterclockwise direction.

Replace the red line with a black edge.

Add a leaf at the unused end of the line, as shown below.

The two leaves are adjacent.

Step 3:

Start at the corner where the previous leaf was placed. Draw a red line [perpendicular](#) to the black line bumping into that corner.

Extend the red line in the [counterclockwise](#) direction.

Draw a new **square** to share an edge with the previous shape, and a second edge with the red line (see the two arrows below).

Make sure you are adding a square that lines up, not a rectangle.

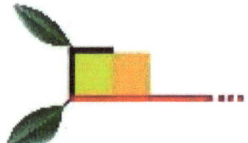

Melt all the present shapes into one structure. Notice there are no squares visible anymore, shapes have blended into a rectangle for now.

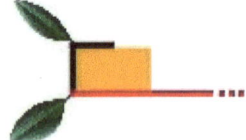

Remove the red line and replace it with a black line. Don't draw past the edge of the shape.

Place a new leaf at the end of the black line.

Step 4:

Jump back to Step 3 and repeat the instructions on your existing shapes. The sequence below shows the results.

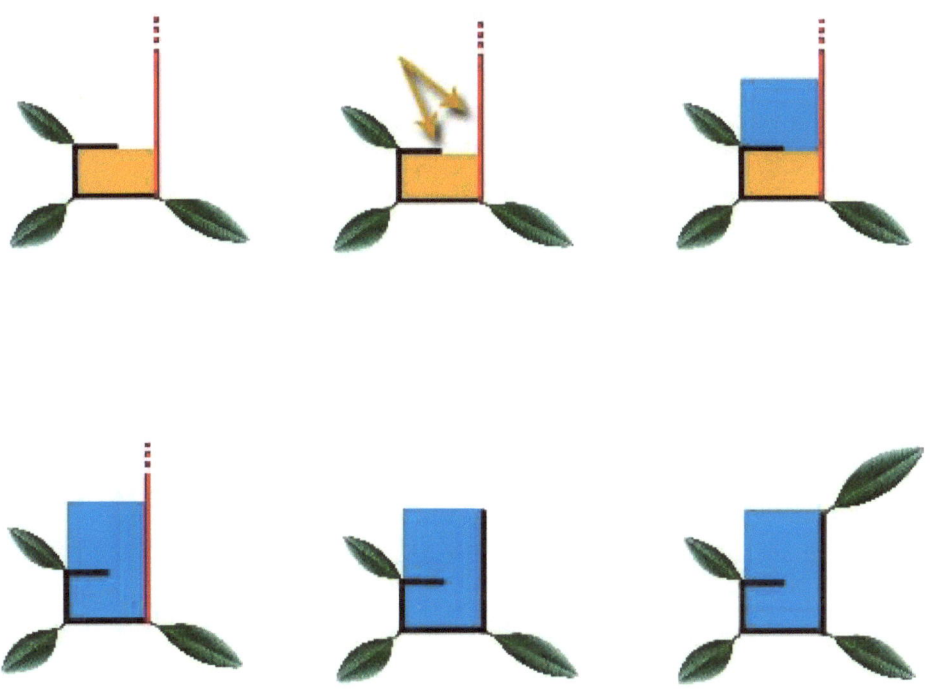

Step 5:

Jump back to Step 3 and repeat the instructions on the existing structure. The following sequence shows the results.

Step 6:

Jump back to Step 3 and repeat the instructions on the current construction. The following sequence shows the results.

Step 7:

After several steps, each looping back to Step 3, the squares are getting larger and the leaves become further apart.

The image below shows how the algorithm progressed after 3 more rounds.

Read the unit below to find out how to get the numerical Fibonacci sequence using the squares.

Test your understanding…
Navigate to *Link 12 at http://www.sharpseries.ca/em/r1.html* and download a printable template and complete the Fibonacci pattern on your own. Shade in squares and follow the algorithm step-by-step. Compare your solution with the one provided on page 2 of that document.

Watch a video...
Navigate to *Link 13 at http://www.sharpseries.ca/em/r1.html* to watch an animation of the Fibonacci pattern evolving the squares and positioning the leaves.

Building the Equivalent Numerical Sequence

The structures in the previous section can be translated into a series of numbers known as the **Fibonacci Sequence**. To construct the number series, begin with the inner most square that started the arrangement. Assume the size of the edge of this square is **1**. Imagine this size is measured in some sort of a **unit** (shown below) like m, cm, mm, etc.

The numerical sequence so far is: 1, ...

Let's calculate measurements for all the black edges. Indicate already calculated edges with dotted lines to track your progress.

The next black line is easy, it belongs to the same square as the first dotted line, thus it is also **1 unit**.

The numerical sequence so far is: 1, 1, ...

The next dotted edge covers the first two identical squares, each 1 unit, together that makes **2 units**. We add this number to the sequence over a dotted line as well.

The numerical sequence so far is: 1, 1, 2, ...

The black line continues to turn [counterclockwise](#) along 2 squares (see arrows).

The numerical sequence so far is: 1, 1, 2, 3, ...

The first square has an edge size equal to 1 unit. The second is 2 units. Together that makes **3 units** which gets assigned to the dotted line. Notice the value is the sum of the last two lines values: 2 + 1.

The next black edge (below) is made of 2 units and another 3 from the two distinct squares it covers. This is in total **5 units**. Notice again this edge's measurements equal the sum of the two previous numbers: 3 + 2.

The numerical sequence is: 1, 1, 2, 3, 5, ...

Upon continuing the calculations, the following pattern emerges: *1, 1, 2, 3, 5, 8, 13, 21, 34, ...* and so on (see below). This numerical pattern is known as the **Fibonacci Sequence**.

The numerical sequence is: 1, 1, 2, 3, 5, 8, 13, 21, 34, ...

The distances between leaves in this algorithm have a growing rhythm. This rhythm is also a ratio. The distances between pairs of leaves are predictable. Each distance equals to the distance of the previous edge plus the one before in the sequence.

Let's examine how the measurements of the distances behave, based on the figure above.

Edge	Previous Edge	The Edge Before That	How They Relate *first column = second column + third column*
34	21	13	34 = 21 + 13
21	13	8	21 = 13 + 8
13	8	5	13 = 8 + 5
8	5	3	8 = 5 + 3
5	3	2	5 = 3 + 2
3	2	1	3 = 2 + 1
2	1	1	2 = 1 + 1

Numbers in the sequence divided by their previous elements are all approximations of the Golden Ratio = **1.6180339887…** Use the table above to generate such divisions: 34/21, 21/13, 13/8, 8/5, 5/3 and 3/2. None of these divisions, or rather ratios, are precise consistently. Each division may yield a slightly different number. Hence, they can only be approximations of the Golden Ratio.

Building the Fibonacci Spiral

An additional concept springs from the above structure: **The Fibonacci Spiral**. To construct the Spiral, draw a quarter of a circle inside every square. Imagine the center of all the circles pointing inwards to where the algorithm started, as shown by the arrows below.

89

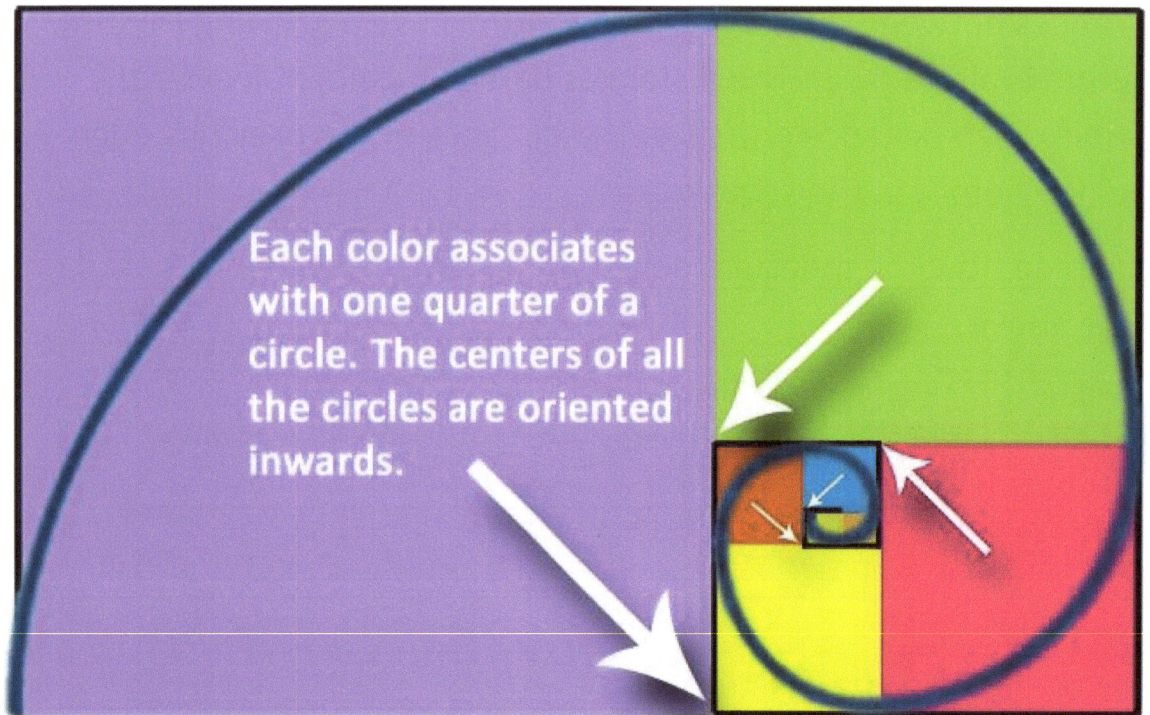

When complete, the quarter circles fit fairly well together into one continuous curve.

Spirals built on the Fibonacci principle are not considered to be perfect. The **Fibonacci Spiral** is nearly the same as the **Golden Spiral**, but not identical, which is why it is commonly known as its approximation.

The **Golden Spiral** is more accurate and is drawn using a different technique using logarithms. Logarithms are an advanced topic, not covered in this volume. For now, think of the two spirals - Golden and Fibonacci - as roughly the same.

Test your understanding...

Use the solution on **page 2** from the earlier exercise: *Link 12 at http://www.sharpseries.ca/em/r1.html*

and a mathematical compass to draw the quarter circles that make the spiral.

Application of Fibonacci in Sciences

As interesting as these number sequences are, one wonders if they have a use in real-life and technologies. A few examples of applications for the Fibonacci sequences outside of Mathematics are:

- **Physics:** Fibonacci Anyons are quasiparticles that may have an important role in the current research and future design of Quantum Computers. These particles can be found in certain highly exotic non-Abelian quantum Hall fluids.
- **Biology:** Codon frequencies in the human genome are positions around 2 fractal-like elements, strongly linked to the golden ratio.
- **Botany:** The Phyllotaxy of various plants such as pine cones, artichokes, Aloe Vera, etc., arrange around their stems on a Fibonacci/Golden ratio.
- **Music:** Many classical music composers are thought to have introduced mathematical patterns such as the Golden Ratio into their compositions to make the music more pleasant.
- **Computer Science**: Computer generated imagery of plants rely on placement of leaves in a Fibonacci sequence to make plants and trees look as realistic as possible.

Take a Quiz on the Fibonacci Topics

Test your understanding of this chapter with a self-grading online quiz. Click the "View your score" link immediately after submitting your answers.

Your response has been recorded.

View your score
Submit another response

Quiz submissions are anonymous. There is no limit to the number of times you can take the quiz.

	Link 14 at http://www.sharpseries.ca/em/r1.html

	Additional Resources, Explore Fibonacci Further
YouTube Video	*The magic of Fibonacci numbers* *Ted Talk: Arthur Benjamin* *Link 15 at http://www.sharpseries.ca/em/r1.html* *(unaffiliated resource)*
Khan Academy Lecture, Video	*The Golden Ratio* *Link 16 at http://www.sharpseries.ca/em/r1.html* *(unaffiliated resource)*
PubMed Abstract Article	*Codon populations in single-stranded whole human genome DNA Are fractal and fine-tuned by the Golden Ratio* *Link 17 at http://www.sharpseries.ca/em/r1.html* *(unaffiliated resource)*
YouTube Video	What Phi (the Golden Ratio) Sounds Like *Link 18 at http://www.sharpseries.ca/em/r1.html* *(unaffiliated resource)*
Music Article	Five Classical Pieces with the Golden Ratio *Link 19 at http://www.sharpseries.ca/em/r1.html* *(unaffiliated resource)*

Chapter 4: Knot Theory

Funny but true, there is a branch of Mathematics that deals with knots. It's predictably called Knot Theory. Knot Theory is part of a mathematical science called **Topology**.

Trefoil Knots

The Trefoil Knot is one of the simplest that is seen when starting to study knot theory. Even though the knot looks simple, its mathematical formula is quite complicated.

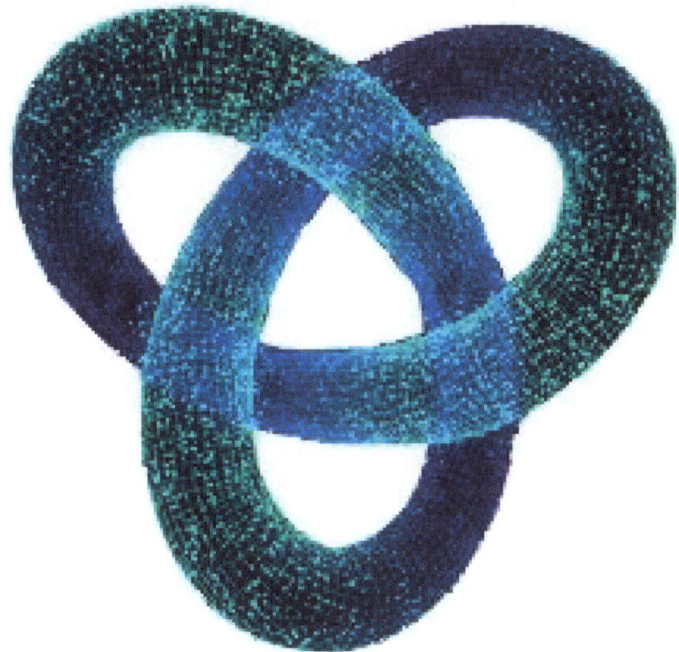

Borromean Rings

Borromean rings happen in threes. There are more complicated arrangements that mimic Borromean systems with an increased number of rings. We'll only examine the three rings arrangement in this book.

Borromean Rings connect in a pattern of overlap. Two rings are weaved by a third, and all rings hold together in the resulting bind. Below you can see a flat 2D version (left) and a 3D (right) version of the rings.

Borromean rings are famous for a simple fact: if you remove one of the rings, the other two naturally fall apart. Yet if the three are together they are completely tied.

Let's follow the [algorithm](#) that binds the rings together.

Linking the rings starts by selecting two of them first. In this example, it's the red and blue. Place the blue ring first, then rest the red ring on its top as shown.

The third ring, the green, is what ties the bunch together by weaving. The green ring must go over the top-most ring and under the ring at the bottom. It weaves its way above and below. You can remember this method by the following sentence: *"the tying ring wants to be higher than the highest ring, and lower than the lowest ring"*.

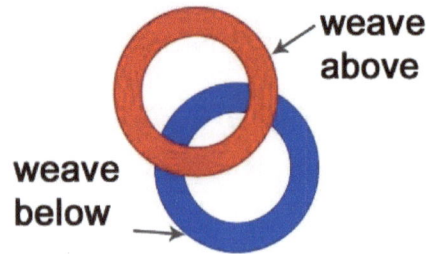

The arrows below show the meandering of the green ring. Notices it always goes above the red, and always under the blue. In other words, higher than the highest and lower than the lowest.

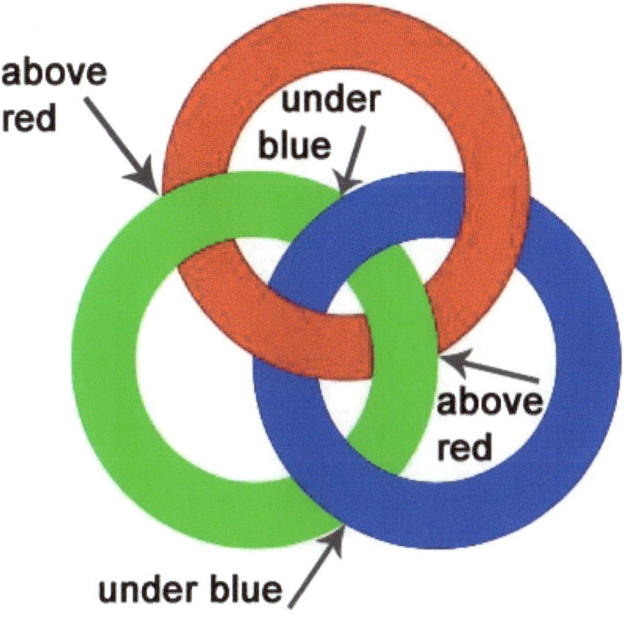

This technique ties the three rings together.

To convince yourself the rings only bind in threes but not in twos, remove the red ring.

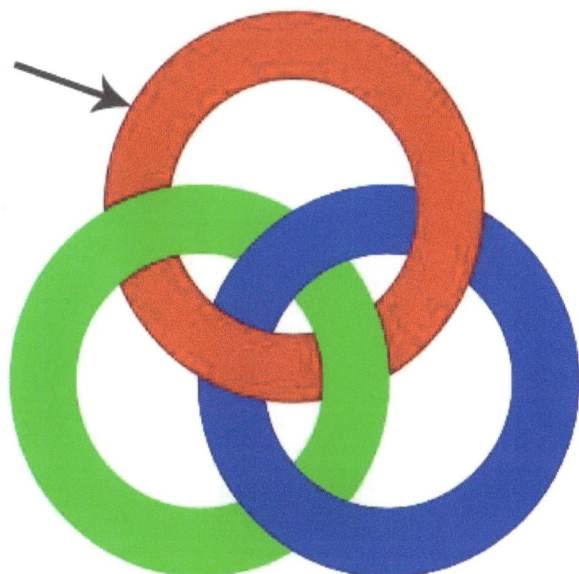

The green and blue rings become disconnected. They will remain one on top of the other without a bind, as shown below.

To test this idea, start over again. This time let the blue ring be the tying one in the method. The green and red rings start the process.

The tying blue ring goes under the red (at the bottom), and over the green (at the top). The result is the same as before.

Experiment with removing any of the rings from here. No matter which is removed from the trio, the other two become disconnected.

	**Test your understanding...**
	Draw three rings, or print an already prepared template for this activity from
	Link 20 at http://www.sharpseries.ca/em/r1.html
	Cut out the rings, and use the third ring to tie the other two together as described in this method or the video below.

	Watch a video about Borromean Rings...
	Watch the activity at:
	Link 21 at http://www.sharpseries.ca/em/r1.html

The technique of binding the rings together is more popular than you may think. The Mathematical name of the most common weave (see below) used to braid hair is called **Borromean Braid**.

The Borromean ring appears as a symbol in earlier civilizations, for example towards the end of the Medieval Age. Rather than circular, it is seen in the triangular variation (below). Triangles replace the circles but connect in the same way. Compare the triangular overlaps with the circular ones. The same colors were used, to show the interlocking mechanism is identical.

Applications of Knot Theory

One of the more interesting facts about Knot Theory is that it is useful in unexpected ways. For example, Knot Theory is helpful in analyzing the ways DNA molecules form their coils. Certain enzymes can twist DNA molecules and make them look like knots. The resulting shapes are difficult to study. Biologists use **Topology** to take advantage of the science already accomplished by Mathematicians so they don't have to reinvent it.

Topology

Topology is a science that evolved out of geometry and amongst other things it studies the attributes of the warped space that is characteristic of knots. There are many Mathematical models on knots, and other topologies, that have been digitalized for easier study.

New sciences like Theoretical Biology rely on digitalized topological models to visualize, zoomed into, twist and tweak their biological equivalents.

An example program that can be used to explore structures in Theoretical Biology is the open-source application **PyMOL**. It is used for molecular visualization. If you are curious what type of insights are provided by such a system, take a look at the video below. You do not need to understand the Biology or the Math to appreciate how interesting this field of study is when combined with the Mathematical counterparts.

Watch a video…
Credit: Daniel Fried. See a video about using PyMOL to explore a molecular structure at: *Link 22 at http://www.sharpseries.ca/em/r1.html (unaffiliated resource)*

Another science, Evolutionary Biology, has found uses for Topology in studying the relationship between base species called genotypes, to their evolved descendants called phenotypes. It seems the genetics between the parent and descendant species have a mathematical relationship that creates paths called evolutionary trajectories, from one to the other. The imaginary space connecting the two species flows its changes, or mutations, through a topological space that obeys mathematical rules. These are fascinating new sciences that await for future researchers like yourself. Being Math-ready will be helpful in assuring participation in these interesting fields.

Take a Quiz on the Knot Theory

Test your understanding of this chapter with a self-grading online quiz. Click the "View your score" link immediately after submitting your answers.

Your response has been recorded.

View your score
Submit another response

Quiz submissions are anonymous. There is no limit to the number of times you can take the quiz.

Link 23 at http://www.sharpseries.ca/em/r1.html

Chapter 5: The Mobius Strip

A Mobius strip is a ribbon that is one-sided and is considered to have paradoxical geometry.

A **Mobius** strip or band, is easy to make by cutting out a paper strip, twisting and gluing the ends together. You can follow along with the process below.

Work along on this project while reading...
Use two strips of colored paper as shown, follow the instructions below and complete the project along while reading the chapter. It will help better understand this experiment.

Let's start with a paper band that has two colors. To make your own, cut two separate bands and tape them together. Pink and white were used in this example. We'll soon find out how a Mobius band connects and what is the curvature of its circumference.

101

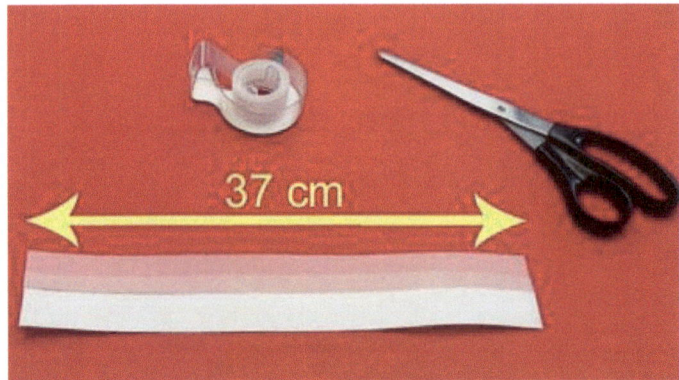

Use a ruler and measure the paper strip before you start. The band here is **37 cm** long. Yours can be whatever length you choose.

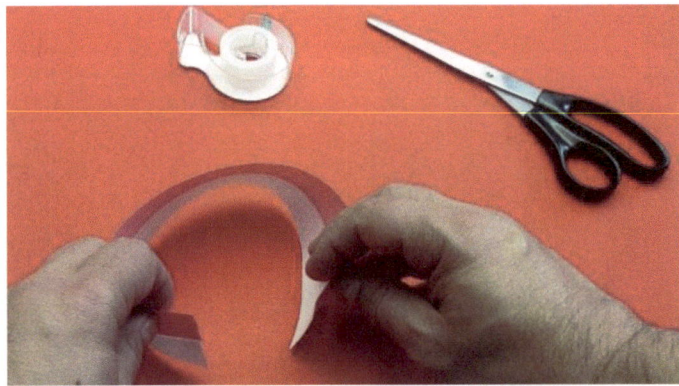

Prepare tape and scissors. Bring the two ends together.

Before you fit the ends, turn the left edge inwards once (see below) for 180°. It means to rotate for half a circle so that white points up and pink points down. Made sure your band only twists once, not more, and that the opposite colors connect: pink to white and white to pink as shown.

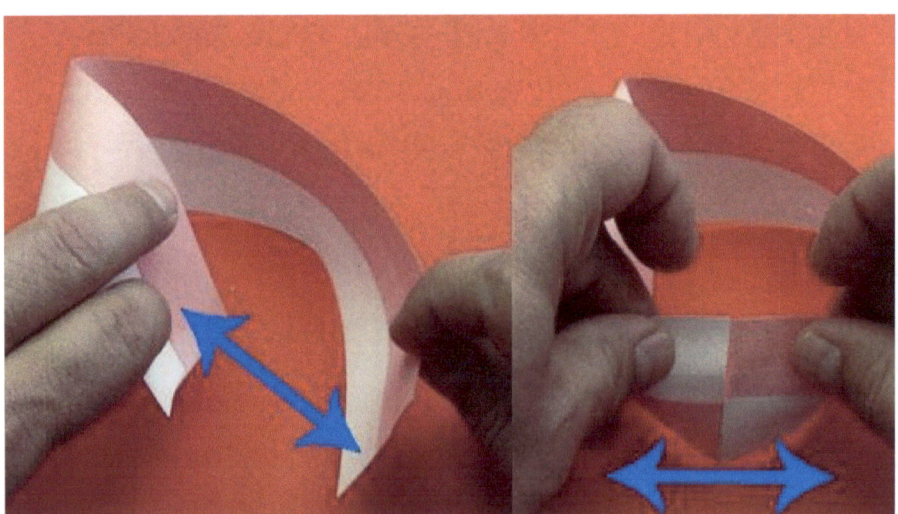

Use tape to secure the two ends together.

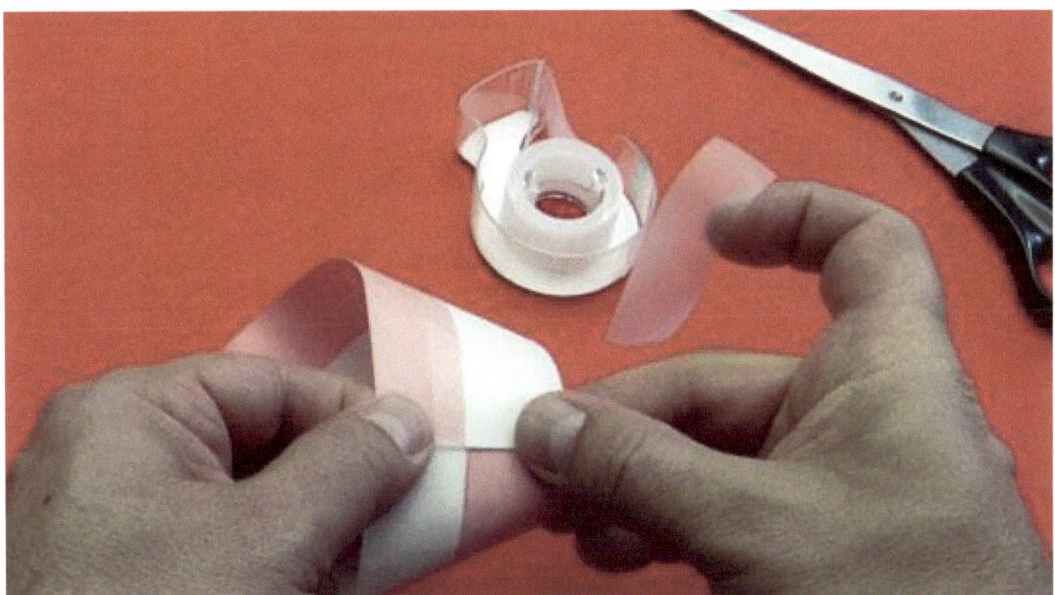

You now have a continuous strip that is known as the **Mobius Band**.

The **Mobius Band** is famous for having a few strange properties. Let's examine them.

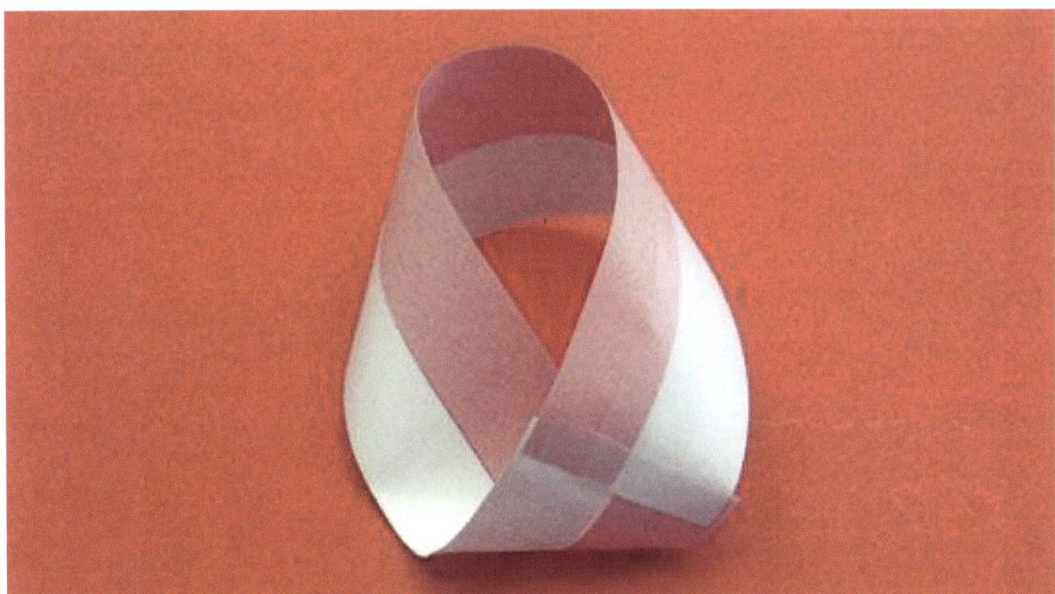

If the band was not fitted in the Mobius way but rather as a circle (without twisting), the circle's circumference would equal the length of the band. In this case that's **37 cm**. For the Mobius strip, drawing a continuous path from one point on its exterior and back to it again, does not equal the circumference of the circle, instead it is longer.

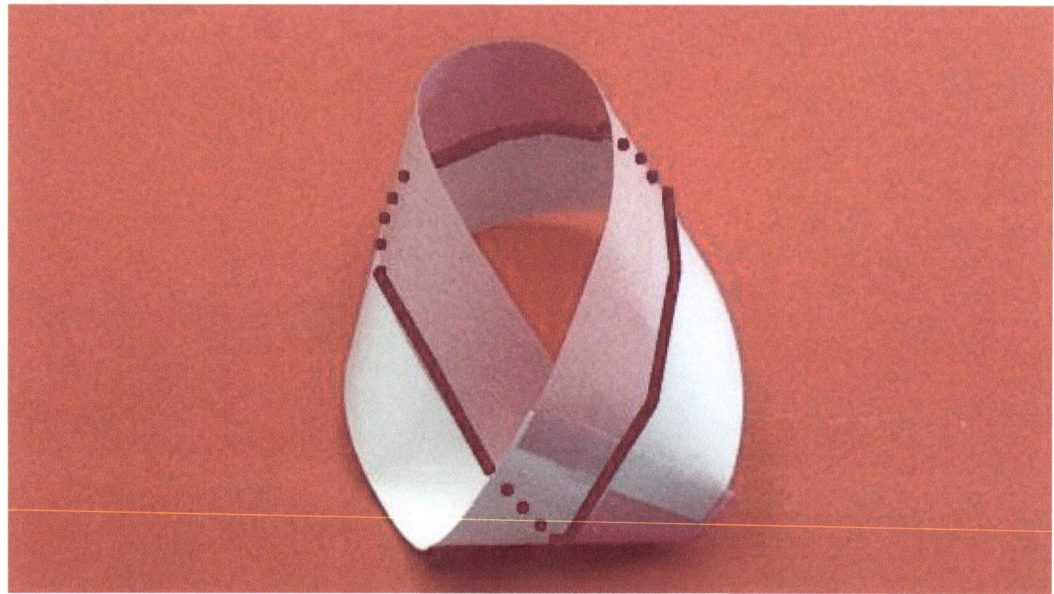

Let's follow the Mobius path from a point and back by tracing your own line. Compute how long the line is. Start tracing where you taped the band together, where the colors swap. We'll use this as a point of reference. This will also be the point we must return to as we trace all the way around.

Keep tracing until you arrive back to where the colors swap, see below, in the middle image. This should have been a full circle and we should meet the point where the red line started. Instead, we find ourselves at the cross-color spot, but what seems to be the other side of the band. So far, **37 cm** were traced, but have not met the start of the line yet.

If you are tracing along, stop for a moment, turn the band around and look where the red line started and where you are. Continue to trace on.

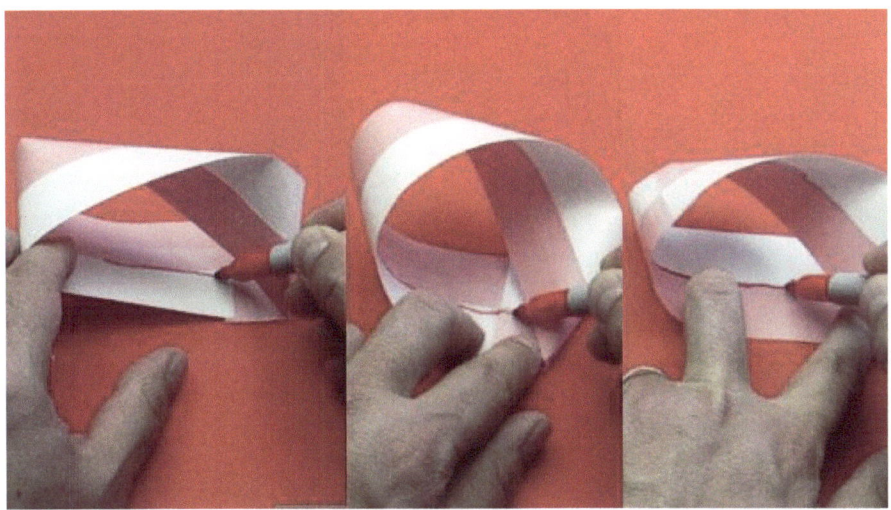

Keep on going until you arrive at a cross-color starting point again. Thus, another full circle is traced for an additional **37 cm**. Together this doubles the length of the red line: 37 cm + 37 cm = **74 cm**, instead of the original 37 cm.

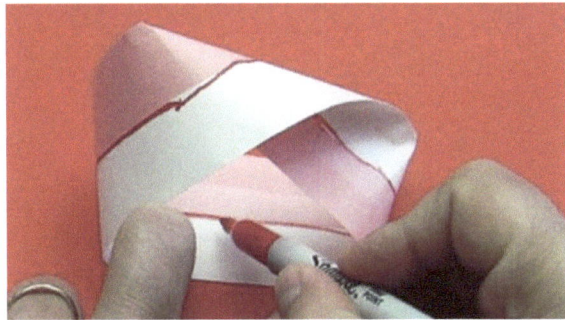

The most important part to notice is that in the process of tracing the band, both the inside and the outside of the Mobius strip were drawn onto without ever having to switch sides.

This is because the inner and outer sides of the Mobius strip are part of one single continuous space, merged together. This is the very detail the Mobius strip is recognized for: **it only has one side**.

The above statement seems paradoxical, as on first look the strip does appear to have two sides. You can even touch the two sides between your fingers, see arrows (A) and (B) below.

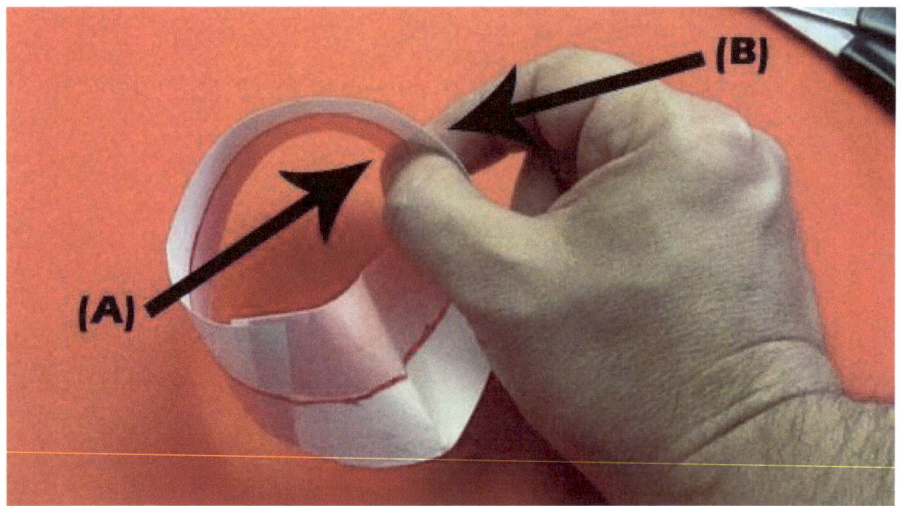

However, in reality, the two arrows are pointing to two positions apart, but still on the same side: the only one side of the band. When tracing a line from point (A) to point (B) it becomes obvious that one would never have to stop and switch sides on the way to the second point. The line would stream continuously undisturbed from one point to the other.

Experiment further with how tracing a line on the Mobius band is different than tracing a circular taped and untwisted strip of paper. To make it even stranger, if you use your scissors to cut along the red line you have just traced, one would expect to get two separate rings, or two separate Mobius strips.

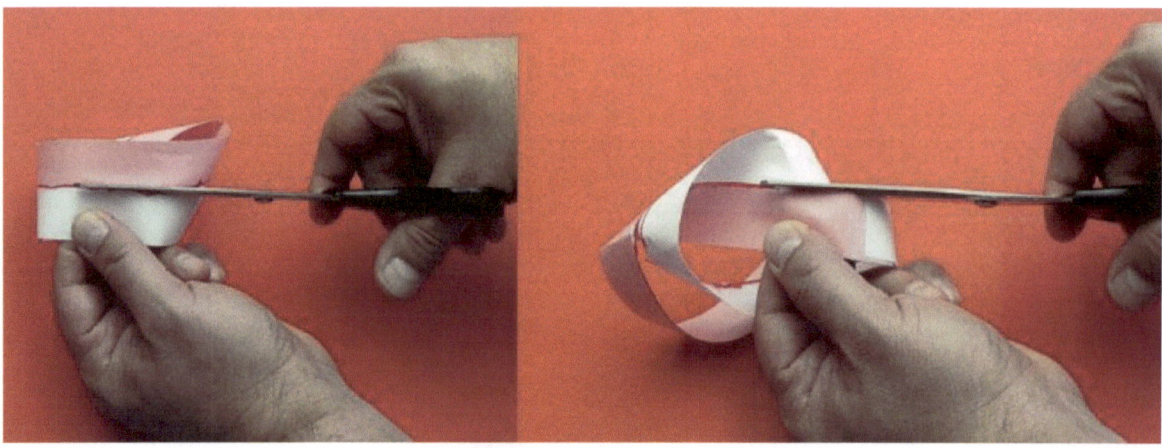

Instead of two bands, the result is the one Mobius strip shown below.

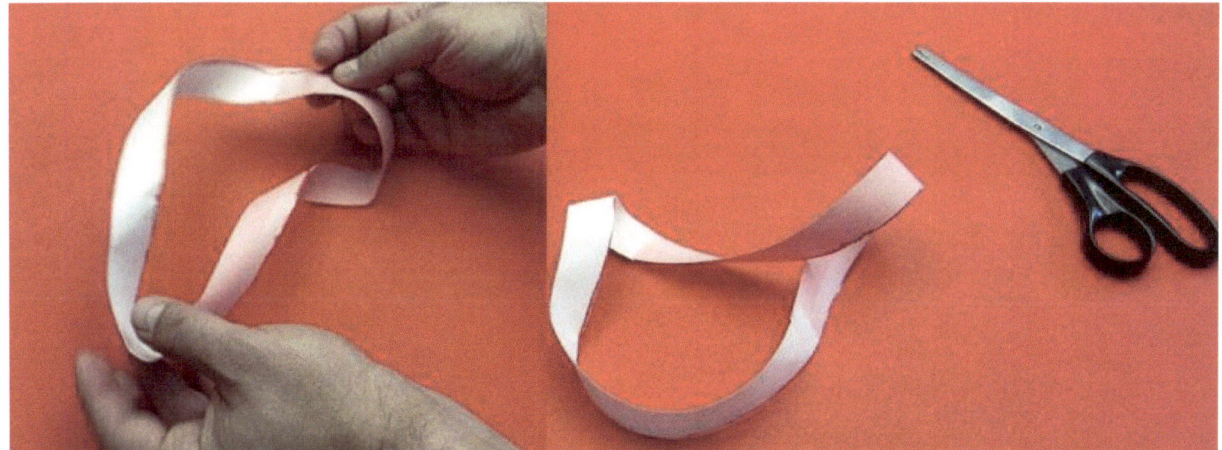

Let's compare the old and new band side by side. The new band (below, left) has two twists, where before there was only one twist (below, right).

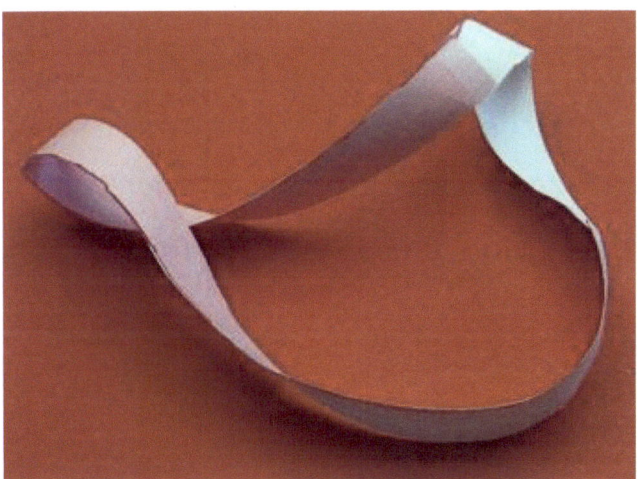

If you thought a pattern emerged, try cutting through the middle of the new band again. If you expected an even larger Mobius strip with a third twist, that is not the case.

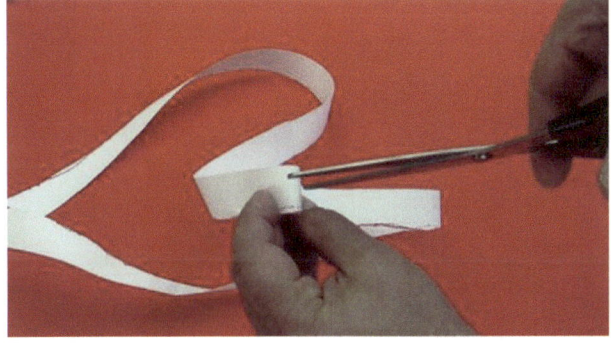
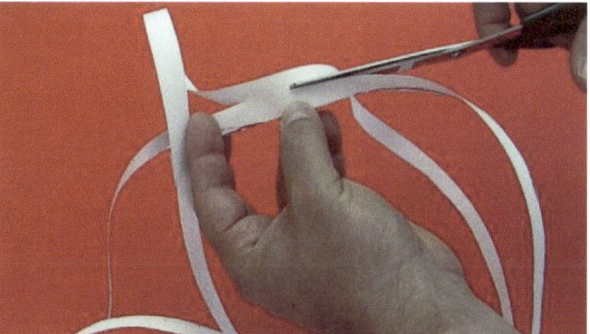

Instead two entangled bands form, identical to the one shown on the left above.

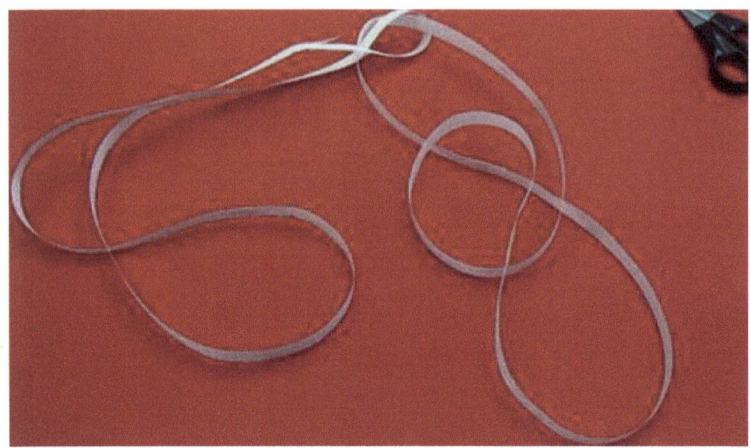

The Mobius Band in the Real World

The Mobius band is frequently seen in art. In a very famous such rendition, ants are traveling on a Mobius Band by artist M. C. Escher.

To see the Mobius Band idea used in architecture, look up images of a recent building called "*Phoenix International Media Center*" in Beijing, China. Its architecture is amazing and futuristic. It is based on the curvature of the Mobius strip.

	Additional Resources, Explore the Mobius Band Further
M.C. Escher Mobius Strip II	*Image Gallery* Link 24 at http://www.sharpseries.ca/em/r1.html (unaffiliated resource)
Arch Daily Article	*Article:* Link 25 at http://www.sharpseries.ca/em/r1.html (unaffiliated resource)
Article	*Phoenix Media Center:* Link 26 at http://www.sharpseries.ca/em/r1.html (unaffiliated resource)

Take a Quiz on the Mobius Strip

Test your understanding of this chapter with a self-grading online quiz. Click the "View your score" link immediately after submitting your answers.

> Your response has been recorded.
>
> View your score
> Submit another response

Quiz submissions are anonymous. There is no limit to the number of times you can take the quiz.

Link 27 at
http://www.sharpseries.ca/em/r1.html

Vocabulary

This section lists definitions of terms. Familiarize yourself with the vocabulary as you read. The digital version of this book permits the use of embedded hyperlinks for each chapter.

Terms do not depend on each other, and appear in alphabetical order.

3D and 2D

3D is pronounced "three dee" and 2D is pronounced "two dee".

3D is present in images that show objects to be as similar as possible to real-life. They have depth, shine, realistic light, and shadows. Sometimes you may not be able to tell if the object is real or not. 3D images are most often created with the help of computers, but some artists have accomplished good results with traditional paintings too.

3D images (left) can be compared to 2D images (right). 2D is flat, with no depth, for example cartoons are such. 2D cartoons may not attempt to be realistic at all.

Adjacent

Two objects are said to be adjacent when they are resting next to each other, or if they are immediate neighbors.

For example, the Fibonacci chapter shows leaves in the corners of rectangles, in that case counting the edges is a quick way to know if two leaves are adjacent: one edge apart is adjacent, two or more edges apart is not.

Algorithm

An algorithm is a method that breaks down some work into steps that anyone can follow. For example, the method for long division is an algorithm.

Some algorithms are designed for people, but most are designed for computers, so the machines can solve problems similar to the way humans do, based on a method that resembles a recipe. To find out more read the Algorithms section in this volume.

Circumference

Imagine any geometric shape and a ball of yarn.

The yarn can be used to trace the outer rim of the shape.

The outer rim of the shape is its circumference.

After wrapping the yarn around the shape, the yarn is cut and measured using a ruler. The length of the yarn measures the circumference of the shape.

Shapes can have straight edges, can be curved, or a combination of both. In each case the outer rim, shown in red below, is the circumference.

Another word for circumference is [perimeter](). The word "circumference" is most often used to describe curved shapes such as circles and ovals.

Clockwise and Counterclockwise

An analog clock is typically round and has two (or three) arms that step forward with the passing of time. The arms of the clock show the hour, the minute, and if there is a third arm the seconds as well.

Clockwise means in circular movement that turns **in the same way** as the passing of time on analog clocks (below, left).

Counterclockwise means in circular movement that turns **in the opposite way** to the passing of time on analog clocks (below, right).

The terms clockwise and counterclockwise are frequently used in Math, Chemistry, and Physics.

 **Clockwise:** the same direction as the passing of time on analog clocks.

 **Counterclockwise:** the direction opposite to the passing of time on analog clocks.

Consecutive

Two numbers that follow one after another in a count, listing or sequence, are called consecutive. For example, 3 and 4 are consecutive in a normal count but 3 and 9 are not because they skip a few numbers in between.

Consecutive is often used to describe steps. Consecutive steps are such that happen immediately one after another. For example, if you have steps 1 to 10 happening in this order, steps labeled 2 and 3 are consecutive but 1 and 7 are out of order. Step 7 doesn't happen immediately after step 1, so step 1 and 7 cannot be consecutive.

Cube

A cube is a geometric shape that is drawn like a box. Its every face is a square. Remember also that boxes with rectangular faces are not cubes.

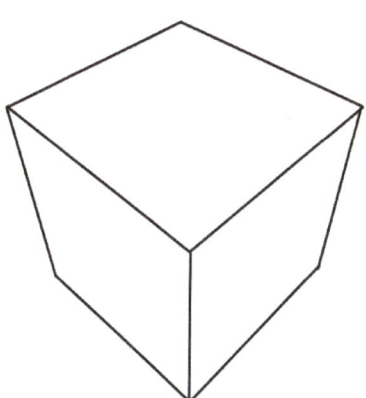

Equilateral

Equilateral means all sides are equal. The black triangle below (left) is equilateral because all its edges are the same length. The red triangle (right) is not equilateral because its edges have different lengths.

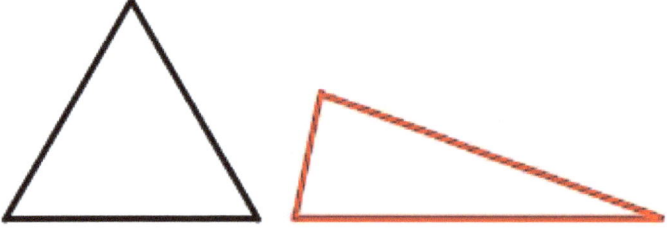

Finite

Finite is the opposite of Infinite. It means there is a limited amount of a thing, and the amount may be counted in a specific time.

Even if a number is large enough that it may not be counted in one's life time, it is still considered to be finite nevertheless. Thus, all numbers that are specific, no matter how large, are finite, even if they are so far along the number line that we can no longer envision them. Such numbers are called **Transfinite** numbers.

Infinite

Infinite is an idea. It means a something that is without an end. When it relates to time, it means forever.

The idea of infinite can be based on several concepts. For example, let's use the number line and the following questions:

- Where does the number line end?
- What is the last possible number on the number line?

The answer is the number line does not have an end, because for any number, the next number in the count can be easily imagined. An addition by one can always be performed, just like when counting.

To demonstrate this, see below a large, astronomical number. The dots in the middle are suggesting there are a lot more digits in that area, but we're not going to list them here. All you need to do is focus on the last one or two digits on the right side of the number (in blue). Count by one from there to figure out the next number.

3,958,445,789,... ,681,235

+1

3,958,445,789,... ,681,236

There is nothing to stop you from counting further, no matter how large a number you have. Such an extremely large number that seems infinite but it isn't, is also known as **transfinite**.

Even transfinite numbers allow you to keep counting past them as far as you want, or have time to.

Another way to say the number line never has to stop is to say it is **Infinite**. Mathematicians use the idea of the infinite frequently, you will see it as this symbol:

The infinite symbol is easy to remember; it looks like a pair of glasses, or an elongated eight turned on its side.

Intersection

The term intersection in Math is similar to what we know from real life. For example, take these two roads that meet in the middle.

Let's mark one road with a red shade and the second road with blue.

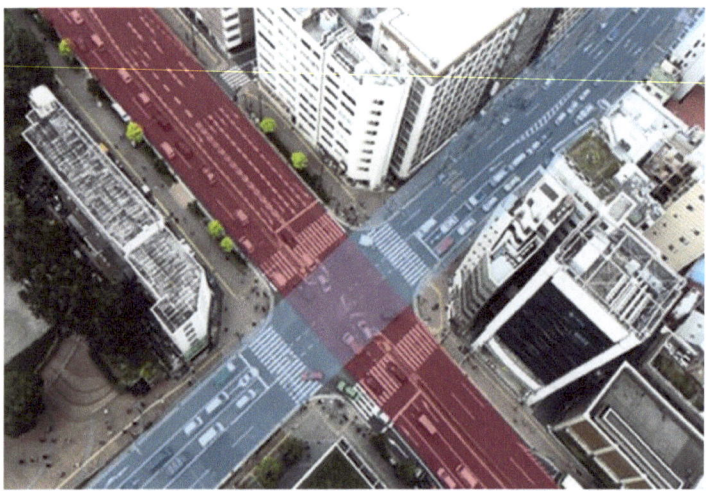

The blue and red roads cross each other and overlap. The overlap is marked with a yellow square below. The yellow area is called an **intersection**.

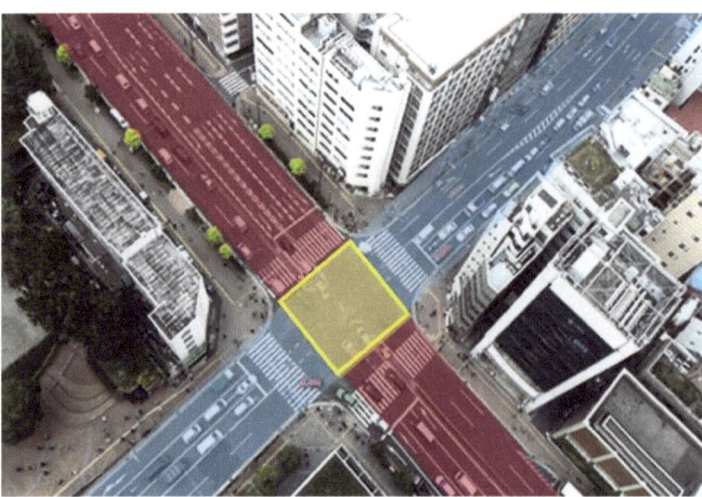

All cars that travel along the red road must go through the yellow square, and the same is true for cars traveling along the blue road. The yellow square is present on both roads equally and indicate where the two roads meet, or intersect. Mathematical expressions relating to intersections are based on the same principle.

An intersection can belong to two geometrical elements, like two roads that run into each other and share a common area. Alternately, intersections can belong to a group of things that can be counted. For instance, the two numerical regions: all numbers between 5 and 10, and all numbers between 9 and 15 intersect in the 9 to 10 range.

Shapes can intersect in a tidy form like the yellow square between the two roads earlier, or can intersect in irregularly shaped zones like the yellow splotch where the star and the cloud overlap below.

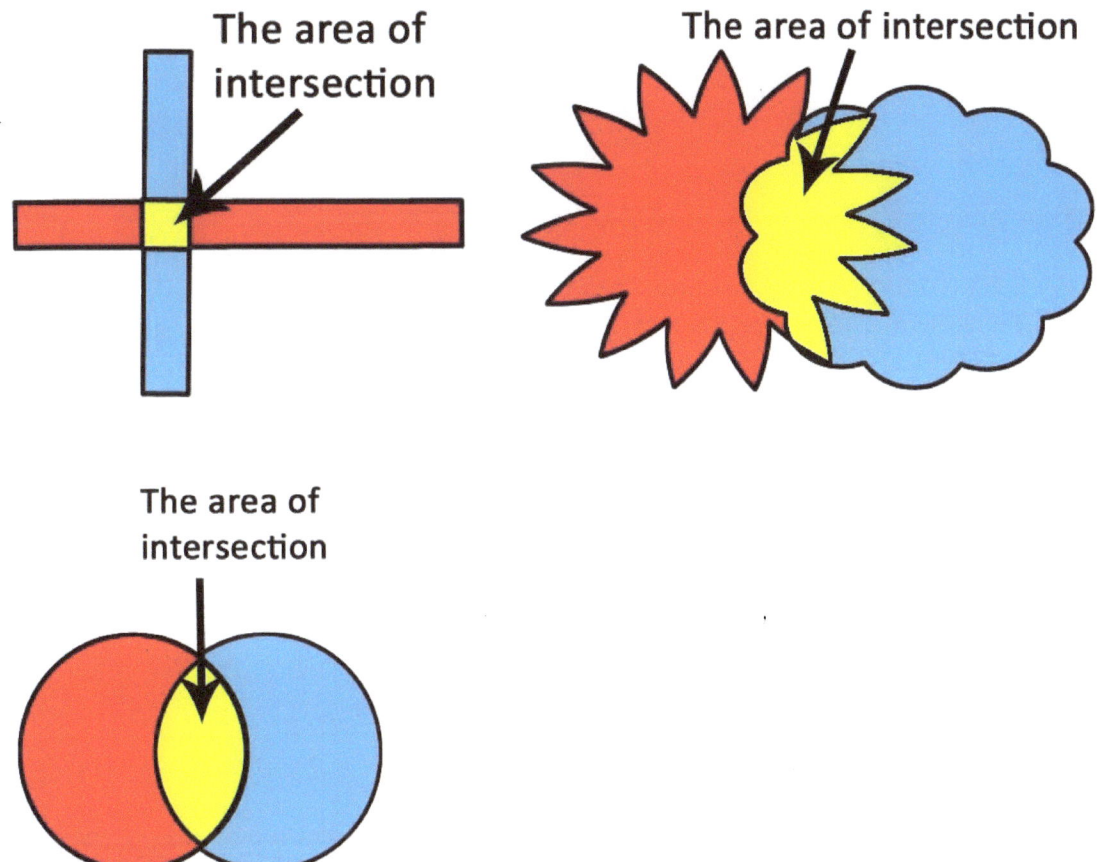

Intersections can be non-geometrical, in which case they represent groups of items like numbers, or anything that can be counted.

Intersections are important when sorting sets of items based on differences or commonalities. More on how to organize sets is described in the [Set Theory](#) chapter.

Microscopic

Microscopic is a size smaller than can be seen by the naked eye. To see such sizes, you require a lens or microscope that can magnify the image.

Examples of microscopic things are: bacteria, fungi spores, cells, and molecules.

Paradoxical

A paradoxical statement is one that seems to make sense initially, but after thinking it through, it leads to a conclusion that contradicts itself, thus not making sense at all. Here are a few examples.

"Impossible is not a word in my vocabulary."

If the word impossible is not part of the person's vocabulary, how then did it get used in a sentence by this person?

A few paradoxical statements and stories have become famous over time. Stories tend to state rules that make some situation impossible. See an example below.

> *There is a village where everyone has hair, and it has only one barber. The village has a law that everyone must follow: all must have their hair cut in the village. If they are cutting their own hair, then they are not permitted to go to the barber. If the hair is not cut by one's self, only the barber is allowed to cut someone else's hair. Can this village exist?*

The answer is that the situation described is paradoxical, and therefore cannot exist. This is because you must ask the question: who cuts the barber's hair? Here is why:

- If the barber cuts his own hair then he is not allowed to go to the barber, but he is the barber so he is already there when the hair is being cut.
- If someone else cuts the barber's hair, then that person is the barber, in which case this barber is not the barber.

The village therefore cannot exist because it offers a paradoxical set of rules. The rules sound suitable initially, but after thinking about it, turns out they cannot all be applied.

Perimeter

The edges of a shape can be measured with either a ruler for straight lines, or with the help of yarn, beads, etc., for curved lines. When adding the length of all the edges together the resulting number is called the length of the Perimeter.

The Perimeter is the path one would walk without stepping off the edges. The path can start from any point on the rim. The direction of movement must always be clock-wise, or always be counterclockwise, so changing directions is not allowed. The path ends on returning to the very point it started from.

See also [circumference](), the perimeter of curved shapes.

Perpendicular

Two lines are called perpendicular to one another if you can precisely fit the corner of a rectangle between them.

Below are examples of perpendicular lines (left). In each case, you can fit a rectangle's corner exactly between the two lines (right).

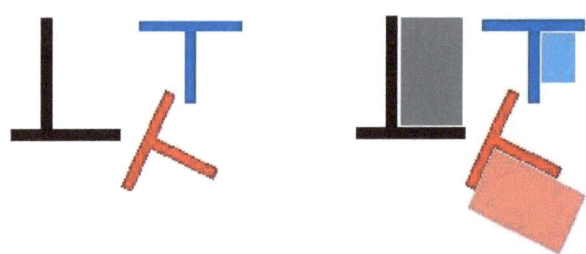

If the lines are not perpendicular (left, below), it is impossible to fit the corner of a rectangle exactly in the space created between the two lines (right, below).

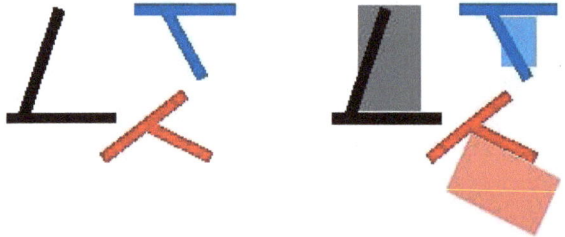

There are many other ways to describe perpendiculars, for example, the angle between perpendicular lines is always 90°.

It can also be said that the two angles on each side of a perpendicular line are equal.

Pyramid

A pyramid is a 3D shape that has a one vertex as a tip, and an edged base that is a polygon. In other words, the base cannot be a circle, oval or curved in any way. Consequently, every face of a pyramid, other than the base, must be a triangle as it connects the tip to the sides of a base edge.

Below are two examples. The pyramid on the left has a square base, and the base of the pyramid on the right is a triangle.

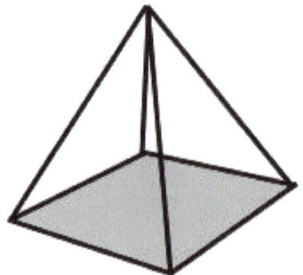 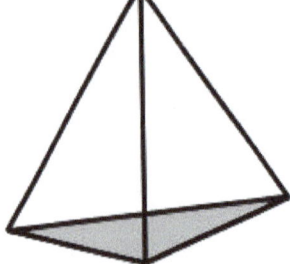

If the base is precisely a circle, then the object would be a cone instead of a pyramid.

See also the definition for [Tetrahedron](), which is any 3D object that has 4 faces. Notice that after accounting for the one face that is the tetrahedron's base, you are left with 3 possible faces for the sides and that forces the tetrahedron base to always be a triangle.

Ratio

Ratios are the same as the two numbers in a division. When describing a ratio, you can refer to the numbers divided, or the result of the division. Ratios are sometimes also called **proportions**.

The idea of a ratio is to show how many times a number fits into another. This can be in whole numbers when the fit is exact, i.e. a whole number: for example, 2 for fitting exactly twice. The fit can be in fractions or decimal numbers for partial fits, for example, 2.5 is for fitting twice and a half times.

Several expressions have evolved over time in the way we talk about ratios. These are simply expressions that can be easily learned.

Following are a few ways of describing the following ratio: **8 divided by 4 is 2**:

i. "**The ratio of 8 to 4 is 2**". The meaning is: 4 fits twice in 8.

ii. "**The ratio of 8 to 4 is 2 to 1**". Each number in the first part is paired with a number in the second part of the sentence. Here 8 and 2 go together (red arrow) and 4 and 1 go together (blue arrow).

The ratio of 8 to 4 is 2 to 1.

The meaning for the paring is to compare that 2 is to 8 (red) as 1 is to 4 (blue).

iii. **"The ratio 8:4 is 2:1"**. This sentence has the same explanation as above. It replaces the word "to" with the colon ":".

The ratio 8:4 is 2:1.

While it is still true that 2 is to 8 (red) as 1 to 4 (blue) like before, it is also true to say: 8 is to 4 (first red and first blue) as 2 is to 1 (second red and second blue).

Most of the time, and for this book as well, all you need to know is that a ratio is the result of a division. We say the ratio 8 to 4 is 2, because 8/4 = 2. If two sets of numbers have the same division result, then they have the same ratio, in other words the sets of numbers are **proportionate**.

Let's explore how geometries compare ratios, or proportions. There are two flower sets in the images below. The ones in the first column, and the ones in the second column. The two columns (large vs. small flowers) are either going to be proportionate (same ratio) or not.

In the example above, the first column has flowers of the same ratios as the ones in the second column. The second column looks like the first column have been miniaturized all in one step together. The flowers were shrunk by the same amount, or in other words, have the **same ratio**, the **same proportions**.

The second column would look differently if it was not proportional to the first column. Each flower in the second column has been shrunk by a different amount. The orange was shrunk further than the magenta flower. The two columns are **not the same ratio**; you can also say they are **disproportionate**.

Spiral

A spiral is a geometric form that looks like a coil, or winded spring.

There are several types of spirals. The difference between them is how much the inner, central part of the coil is winding when compared to the outer parts. Spirals can be equally strained all over (left), or with more strain in the center and more relaxed parts outwards (right).

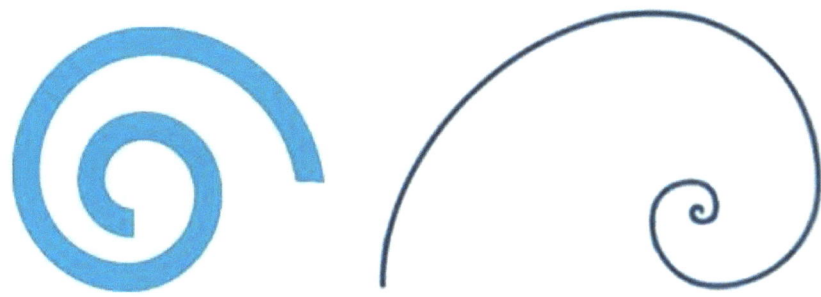

The spiral on the left is known as an Archimedean spiral. The one on the right is a version of the Fibonacci/Golden spiral.

Tetrahedron

Tetrahedron is a 3D geometrical shape. "Tetra" in the name means four, to indicate that the shape must always have 4 faces. A tetrahedron uses one face for the base, and thus 3 out of the 4 can be used for the sides. The only way 3 faces can rest on the base, is if each gets its own base edge "to live on", thus the base must have precisely 3 edges, which makes it a triangle, and the 3D shape overall a triangular pyramid.

Additionally, all sides of the tetrahedron (faces other than the base) must be triangles as well since they are made of a tip and a base edge.

If the base of the pyramid is more complex than a triangle, the pyramid then forms more than 4 side faces and can no longer be considered to be a tetrahedron.

A tetrahedron is shown below from two different angles. A flower was placed in a position you can track when the shape is turned around. Based on the position of the flower let's call the left image the back. The version in the middle, let's call this the front, and is what we see after turning the pyramid around so that the flower can face us. The rightmost shape shows how the middle version would look like if it was transparent and you could see the back edge while looking from the front. The hidden edge is shown as an interrupted line.

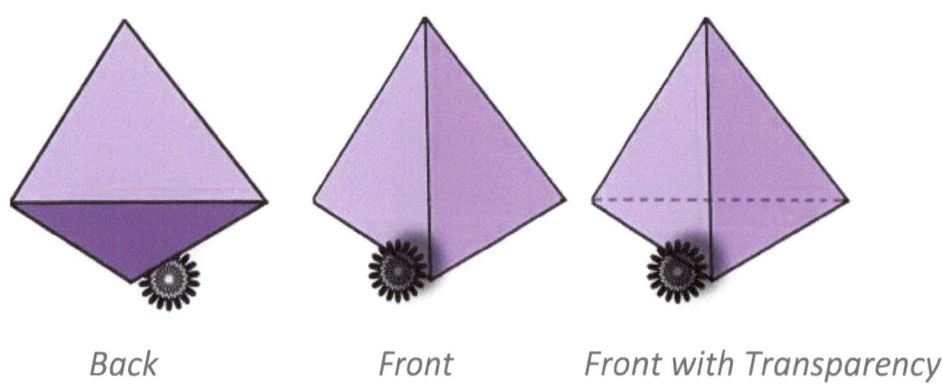

Back *Front* *Front with Transparency*

Many Mathematicians while sketching shapes for 3D geometry, like to show the hidden edges with a dotted or interrupted line. This better describes what's going on with a shape even in the areas that may not be directly visible from the given angle.

Errata and Feedback

This book's errata can be found at:

http://www.sharpseries.ca/earlyMath/errata.html

Comments and suggestions for future editions are welcome.